Contract Administration: Technology and Practice in Europe

Prepared by the Scanning Team:

David O. Cox
FHWA

Keith R. Molenaar
University of Colorado

James J. Ernzen
Arizona State University

Gregory Henk
Flatiron Structures Co., LLC

Tanya C. Matthews
Design-Build Institute of America

Nancy Smith
Nossaman, Gunther, Knox & Elliott

Alex Levy
FHWA

Ronald C. Williams
Arizona DOT

Frank Gee
Virginia DOT

Jeffrey Kolb
FHWA

Len Sanderson
North Carolina DOT

Gary C. Whited
Wisconsin DOT

John W. Wight
HNTB Corporation

Gerald Yakowenko
FHWA

and

American Trade Initiatives, Inc.
&
Avalon Integrated Services, Inc.

for the

Federal Highway Administration
U.S. Department of Transportation

and

The American Association of State Highway and Transportation Officials

and

The National Cooperative Highway Research Program
(Panel 20-36)
of the Transportation Research Board

October 2002

Technical Report Documentation Page

1. Report No.	2. Government Accession No.	3. Recipient's Catalog No.
FHWA-PL-02-0xx		

4. Title and Subtitle	5. Report Date
Contract Administration: Technology and Practice in Europe	October 2002
	6. Performing Organization Code
7. Author(s) David O. Cox, Keith R. Molenaar, James J. Ernzen, Gregory Henk, Tanya C. Matthews, Nancy Smith, Ronald C. Williams, Frank Gee, Jeffrey Kolb, Len Sanderson, Gary C. Whited, John W. Wight, Gerald Yakowenko	8. Performing Organization Report No.
9. Performing Organization Name and Address American Trade Initiatives P.O. Box 8228 Alexandria, VA 22306-8228	10. Work Unit No.(TRAIS)
	11. Contract or Grant No. DTFH61-99-C-0005
12. Sponsoring Agency Name and Address Office of International Programs Office of Policy Federal Highway Administration U.S. Department of Transportation	13. Type of Report and Period Covered
	14. Sponsoring Agency Code

15. Supplementary Notes
FHWA COTR: Donald W. Symmes, Office of International Programs

16. Abstract
In June 2001 a team comprising federal, State, contracting, legal, and academic representatives traveled to Portugal, the Netherlands, France, and England to investigate and document alternative contract administration procedures for possible implementation in the United States.

The scan team discovered that European highway agencies appear to be better exploiting the efficiencies and resources that the private sector offers, through the use of innovative financing, alternative contracting techniques, design-build, concessions, performance contracting, and active asset management. European agencies have created contracts that focus on the users, while seeking to allocate risk appropriately and establish an atmosphere of trust in the implementation of procedures. The United States can directly and immediately employ many European procedures to help cope with its most urgent transportation needs. The report discusses these European techniques in terms of procurement, contract types, and payment mechanisms.

17. Key Words	18. Distribution Statement
Best-value selection, performance specifications, design-build, shadow tolls, public-private partnerships, concessions, design-build-operate-maintain	No restrictions. This document is available to the public from the Office of International Programs FHWA-HPIP, Room 3325 US Dept. of Transportation Washington, DC 20590 international@fhwa.dot.gov www.international.fhwa.dot.gov

19. Security Classif. (of this report)	20. Security Classif. (of this page)	21. No. of Pages	22. Price
Unclassified	Unclassified	xx	Free

Form DOT F 1700.7 (8-72) Reproduction of completed page authorized

FHWA INTERNATIONAL TECHNOLOGY EXCHANGE PROGRAMS

The FHWA's international programs focus on meeting the growing demands of its partners at the Federal, State, and local levels for access to information on state-of-the-art technology and the best practices used worldwide. While the FHWA is considered a world leader in highway transportation, the domestic highway community is very interested in the advanced technologies being developed by other countries, as well as innovative organizational and financing techniques used by the FHWA's international counterparts.

INTERNATIONAL TECHNOLOGY SCANNING PROGRAM

The International Technology Scanning Program accesses and evaluates foreign technologies and innovations that could significantly benefit U.S. highway transportation systems. Access to foreign innovations is strengthened by U.S. participation in the technical committees of international highway organizations and through bilateral technical exchange agreements with selected nations. The program has undertaken cooperatives with the American Association of State Highway Transportation Officials and its Select Committee on International Activities, and the Transportation Research Board's National Highway Research Cooperative Program (Panel 20-36), the private sector, and academia.

Priority topic areas are jointly determined by the FHWA and its partners. Teams of specialists in the specific areas of expertise being investigated are formed and sent to countries where significant advances and innovations have been made in technology, management practices, organizational structure, program delivery, and financing. Teams usually include Federal and State highway officials, private sector and industry association representatives, as well as members of the academic community.

The FHWA has organized more than 50 of these reviews and disseminated results nationwide. Topics have encompassed pavements, bridge construction and maintenance, contracting, intermodal transport, organizational management, winter road maintenance, safety, intelligent transportation systems, planning, and policy. Findings are recommended for follow-up with further research and pilot or demonstration projects to verify adaptability to the United States. Information about the scan findings and results of pilot programs are then disseminated nationally to State and local highway transportation officials and the private sector for implementation.

This program has resulted in significant improvements and savings in road program technologies and practices throughout the United States, particularly in the areas of structures, pavements, safety, and winter road maintenance. Joint research and technology-sharing projects have also been launched with international counterparts, further conserving resources and advancing the state of the art.

For a complete list of International Technology Scanning topics, and to order free copies of the reports, please see list on the facing page.

Website: www.international.fhwa.dot.gov
Email: international@fhwa.dot.gov

FHWA INTERNATIONAL TECHNOLOGY EXCHANGE REPORTS

International TechnologyScanning Program: Bringing Global Innovations to U.S. Highways

Infrastructure

Geotechnical Engineering Practices in Canada and Europe
Geotechnology—Soil Nailing
International Contract Administration Techniques for Quality Enhancement-CATQEST

Pavements

European Asphalt Technology
European Concrete Technology
South African Pavement Technology
Highway/Commercial Vehicle Interaction
Recycled Materials in European Highway Environments

Bridges

European Bridge Structures
Asian Bridge Structures
Bridge Maintenance Coatings
European Practices for Bridge Scour and Stream Instability Countermeasures
Advanced Composites in Bridges in Europe and Japan
Steel Bridge Fabrication Technologies in Europe and Japan
Performance of Concrete Segmental and Cable-Stayed Bridges in Europe

Planning and Environment

Sustainable Transportation Practices in Europe
Wildlife Habitat Connectivity Across European Highways
European Right-of-Way and Utilities Best Practices
European Intermodal Programs: Planning, Policy and Technology
National Travel Surveys
Recycled Materials in European Highway Environments
Geometric Design Practices for European Roads

Safety

Pedestrian and Bicycle Safety in England, Germany and the Netherlands
Speed Management and Enforcement Technology: Europe & Australia
Safety Management Practices in Japan, Australia, and New Zealand
Road Safety Audits—Final Report
Road Safety Audits—Case Studies
Innovative Traffic Control Technology & Practice in Europe
Commercial Vehicle Safety Technology & Practice in Europe
Methods and Procedures to Reduce Motorist Delays in European Work Zones

Operations

Freight Transportation: The European Market
Advanced Transportation Technology
European Traffic Monitoring
Traffic Management and Traveler Information Systems
European Winter Service Technology
Snowbreak Forest Book – Highway Snowstorm Countermeasure Manual (*Translated from Japanese*)
European Road Lighting Technologies

Policy & Information

Emerging Models for Delivering Transportation Programs and Services
Acquiring Highway Transportation Information from Abroad—Handbook
Acquiring Highway Transportation Information from Abroad—Final Report
International Guide to Highway Transportation Information

All publications are available on the internet at www.international.fhwa.dot.gov

Abbreviations and Acronyms

AASHTO	American Association of State and Highway Transportation Officials
AMPM	Active management payment mechanism
ARTBA	American Road and Transportation Builder's Association
CATQUEST	1994 Contract Administration Techniques for Quality Enhancement Study Tour
CIRIA	Construction Industry Research and Information Association
DBFO	Design-build-finance-operate
DBOM	Design-build-operate-maintain
DOT	Department of Transportation
EIB	European Investment Bank
EU	European Union
FHWA	Federal Highway Administration
FTA	Federal Transit Administration
ID/IQ	Indefinite delivery/indefinite quantity
IEP	Instituto das Estratas de Portugal, Portuguese Highways Agency
IRS	Internal Revenue Service
ISC	Integrated supply chain
MAC	Managing agent contract
NCHRP	National Cooperative Highway Research Program
NPV	Net present value
PFI	Private finance initiative
PFMAC	Private financing managing agent contract
PPC	Public-Private Comparator
PPP	Public-private partnership
PRIDe	Performance Review Improvement and Delivery group
QA/QC	Quality assurance/quality control
RFP	Request for proposal
RFQ	Request for qualification
RWS	Rijkswaterstaat, Dutch national department in charge of infrastructure, including roads
SHA	State highway agency
SIB	State infrastructure banks
STIP	Scanning Technology Implementation Phase

Contents

Chapter 1: Executive Summary

Many transportation agencies have discovered that traditional highway contract administration procedures and project delivery methods do not meet current demands. In the United States, both federal and State agencies are turning to alternative contracting procedures to accommodate reconstruction and growth. The primary goal for the majority of alternative contracting procedures is to deliver projects faster, without compromising safety or quality or increasing costs. Quite often, increased safety, higher quality, and decreased cost can be achieved, while delivering projects at a faster pace. In the United States, the use of alternative contracting practices has been on the rise since the early 1990s, and the highly publicized success of numerous "mega projects" is encouraging more agencies to experiment with alternative contracting methods. Likewise, numerous European nations are employing alternative contracting methods to meet increasing infrastructure needs. Recognizing the benefits that could result from an international examination of alternative contracting procedures, a diverse team of experts was assembled to research, document, and promote the implementation of best practices found in Europe that might benefit U.S. industry. The Federal Highway Administration (FHWA) and the American Association of State and Highway Transportation Officials (AASHTO) jointly sponsored this study, under the National Cooperative Highway Research Program (NCHRP).

In June of 2001, a team consisting of federal, State, contracting, legal, and academic representatives traveled to Europe to investigate and document alternative contract administration procedures that are employed in Europe to cope with growing transportation needs. Appendix A contains the names, affiliations, and biographic information of the scanning team members. The team traveled to Lisbon, Portugal; The Hague, the Netherlands; Paris, France; and London, England. Additionally, the team met with Swedish transportation officials while in the Netherlands. The ministries of transportation, numerous private-sector contractors, and research organizations involved in contract administration hosted the team. Appendix B lists the names of the organizations and their representatives.

In recent years, the European community has faced a multitude of problems that are similar to those that challenge the U.S. transportation community today. The scan team discovered that European highway agencies appear to be better at exploiting the efficiencies and resources that the private sector offers, through the use of innovative financing, alternative contracting techniques, design-build, concessions, performance contracting, and active asset management. European agencies have created contracts that focus on the users, while seeking to allocate risk appropriately and establish an atmosphere of trust in the implementation of procedures. The United States can directly and immediately employ many European procedures to help cope with its most urgent transportation needs.

DRIVERS OF CHANGE IN EUROPE

Until the late 1980s, for the most part, European methods of contract procurement and administration were very similar to those in the United States. Public transportation agencies retained tight control over the design and construction of the highway systems. Prescriptive specifications and low-bid procurement methods were the public-sector tools of choice for procuring new works in both the United States and Europe. In the late 1980s, European agencies began to make significant changes to contract administration techniques. While various U.S. transportation agencies experimented with alternative contracting methodologies starting about the same time, the European agencies started to use alternative methodologies as their primary contracting methodology for major projects. The scan team quickly realized that the drivers for change in Europe include some of the same problems in the United States today. Some of the most significant drivers of change confronting Europe include:

- Growing infrastructure needs
- Inadequate public funds
- Insufficient and diminishing staff
- Lack of innovation in addressing project needs
- Slow product delivery and delays
- Cost overruns
- Adversarial relationships
- Claims-oriented environments
- Perceived lack of maintenance efficiency
- New European Union (EU) directives
- User frustration
- Political discontent

These problems are certainly not unique to Europe; most U.S. States share some, if not all, of them. (Even though EU members have a different relationship with the EU than U.S. States have with the federal government, the EU Directives are analogous to Title 23 and FHWA's regulations.) This report describes tools and techniques that European transportation agencies and private-sector groups have used to overcome their problems. Many of the tools and techniques can be directly and immediately applied in the United States, if legislative and political environments allow. Other techniques may be valuable in the future or could serve as indicators of future contracting types.

CONTRACTING TECHNIQUES

European transportation agencies are implementing a wide variety of alternative contracting techniques that could have a tremendous impact on the efficiency and effectiveness of contract administration in the United States. The report discusses these techniques in terms of procurement, contract types, and payment mechanisms. Similar to the relationship between the FHWA and State departments of transportation (DOTs), the EU directives establish minimum requirements that must be used by its members for procurement, but individual countries can develop unique contracting techniques that fit distinctive needs.

The most notable difference between European and U.S. procurement methods is that best-value awards are widely used in all types of procurements. Low-bid selection, although still used, is becoming less common. The Europeans have found that best-value selection, using transparent and uniform processes, enhances competition and innovation. In the case of long-term maintenance contract procurements, the business culture and quality are weighted much more significantly than the price and technical portions of the procurement. Short listing is widely used to ensure that all potential proposers are competent technically and meet the owner's other minimum requirements. In cases of public-private ventures and privatization, careful consideration is given to the economic benefits of the procurement. The public-sector transportation agencies have dedicated significant effort to evaluating and assessing best-value proposals, and, in some cases, have significantly changed their organizational structures. Finally, the ministries of transportation visited by the scan team use confidential discussions in their procurement processes much more readily than in the United States. The European agencies provided examples of an increase in design and construction innovation resulting from these discussions in the procurement phase.

This report discusses a number of contract types being used in Europe. The United States is currently employing a number of these techniques, but the scan revealed new techniques that have merit for consideration in the States. Some of the contract types discussed in this report appear in the table below. Specific examples are discussed later in the report.

Contracts Similar to U.S. Methods	Contracts not Currently Used by U.S. Agencies
• Design-Build	• Framework Contracts
• Design-Build-Maintain	• Management Agency Contracting (MAC)
• Design-Build-Operate-Maintain	• Private Finance MAC
• Concessions	• Integrated Supply Chain Management

In summary, all of these types of contracts promote creation of partnerships between the public and private sectors. European agencies are actively working toward development of relationships with the private sector that are based on trust and delegation of responsibility. The contracts discussed in this report provide examples of how some European countries are allocating contractual risk to leverage the efficiency of the private sector to provide benefits to the public.

Certain alternative procurement methods and contracts can combine nontraditional payment mechanisms to optimize their benefits. In many cases, payments are not based on units of work completed, but rather on availability of the product at the end of the project. The private-sector providers are required to finance the cash flow during and after construction. They ultimately receive payments based on factors such as availability (i.e., number of lanes open), quality of performance (i.e., smoothness), and/or safety (a reduction in the number of crashes, measured against a baseline). Disincentives were observed on maintenance contracts, and incentives were readily used for safety.

DESIGN-BUILD

In the countries visited, design-build was observed to be the contracting method of choice for many types of projects, ranging from green-field construction to pure maintenance contracts. Design-build also is an inherent component in concessions and public-private partnerships. In the United Kingdom, the Highways Agency's contracting method of choice is design-build. Design-build contracts are typically awarded on a best-value basis. In the best-value analysis, lifecycle costs are analyzed using net present value (NPV). In the United Kingdom, the Highways Agency indicated that in the early 1990s it carried preliminary designs too far, prior to tendering. The agency has now corrected that error. One area where the Europeans appear to be more advanced than the Americans is in writing outcome (value) specifications. U.S. practitioners are struggling with similar performance specifications. This report includes some tools observed for developing outcome specifications that are directly and immediately applicable to U.S. design-build practices. In Europe, the issue of quality assurance in design-build contracts is primarily dealt with through the use of 5- to 10-year warranties and 30-year concessions. The use of alternative financing, operation, and maintenance, in conjunction with design-build contracts, minimizes the need for owners to perform time-consuming and redundant inspection and testing. The lessons learned on this scan tour include the types of projects suitable for design-build, the use of best-value selection for design-build projects, the need to minimize the level of design in the solicitation, design and construction administration, third-party risks, the use of warranties, and the addition of maintenance and operation to design-build contracts. In summary, the design-build techniques observed in Europe promote a level of partnering and early contractor involvement not yet widely seen in the United States.

PERFORMANCE CONTRACTING

Performance contracting is in its infancy in the U.S. transportation sector, but the tools and techniques are well established in Europe. Performance contracting allows the contractor to employ whatever means it determines are most appropriate (and economical) to satisfy the performance specifications provided by the owner. Performance contracts allow innovation through creative design and construction methods—and are thought to lower the overall price of a given project. Performance contracts necessitate alternative procurement practices with past performance and innovative solutions as major factors in the selection process. Such contracts also are ideal candidates for alternative payment mechanisms, typically using end-product qualities as measurements.

Performance specifications are critical elements of performance contracting. In the Netherlands, the Highways Agency has extensive experience with drafting performance specifications. The Dutch are testing a series of 60 pilot projects to measure performance contracting versus traditional prescriptive methods. They define performance specifications in five levels of requirements that range from road-user wishes to requirements for basic materials and processing. Performance specifications detail both the operating level and minimum condition of the facility at the time it is returned to public ownership.

An area of concern in performance contracting in the United States is quality assurance/quality control (QA/QC). Traditional QA/QC roles and responsibilities in the United States can impede the effectiveness of performance contracting. Performance contracts observed by the scan team placed the responsibility for QC solely with the contractor, and the owner retained only a minimal QA role. Owner QA is built into the process at various "stop" or "control" points on projects. There also are unique processes for penalty points and quality audits in lieu of heavy owner inspection. In one instance, the owner gives the contractor yellow or red cards for quality violations, like a referee in a soccer game. One yellow card is a warning and allows the contractor to correct work while improving its process or fixing the problem. Two yellow cards, or one red card, mean that the contractor must stop work until the violation is remedied.

ALTERNATIVE FINANCING

Many of the alternative financing techniques in use in Europe have the potential to be used in the United States. Two significant differences between the U.S. and European finance processes, however, must be considered. First, the countries visited do not have tax revenue sources dedicated exclusively to transportation needs. This situation means that gasoline taxes and the like are not earmarked for transportation projects, but are deposited into a general fund with other taxes. The general funds provide money for a variety of needs, including transportation projects, but no taxes are specifically dedicated for future transportation projects. The second difference is that European governments do not have the ability to use tax-exempt financing for public transportation projects, as is the case in the United States. Although this means that interest rates are higher for European projects, it also means that such projects are not subject to the management contracting rules applicable to U.S. projects using tax-exempt financing, and makes private financing much more competitive with public financing. For example, in the United Kingdom, the interest differential between publicly guaranteed funds and private funds is sometimes less than 1 percent.

Alternative funding sources in Europe include a combination of bond and bank financing. Private financing is used much more readily than in the United States. In some cases, private financing is used because governments have reached ceilings for public debt; in others cases, it is simply because private financing is a competitive solution. For example, the have Dutch created a toll tunnel project through a limited-liability entity and plan to transfer ownership to the private sector by selling shares of the entity to the public when the tunnel is operating profitably. Meanwhile, in Portugal, concessionaires bid for the rights to maintain and operate existing highways, creating a type of off-balance sheet approach to government funding and even purchase of highway infrastructure.

The scan revealed several alternative financing payment mechanisms. As in the United States, real tolls are in use, but, in some situations, real tolls meet with public and political resistance. Both Portugal and the United Kingdom are experimenting with systems of "shadow tolling". Shadow tolls involve payment of user fees by the government on the basis of the number of vehicles that use the facility, allowing the concessionaire to obtain financing for the project secured by the user fees and based on traffic studies. The user fees are paid on the basis of traditional sampling methods and high-tech count mechanisms that establish the number of vehicles using the facility. This arrangement gives the concessionaire the risk of, and reward for, the number of vehicles using the road. In the United Kingdom, shadow toll arrangements are evolving from a "toll per vehicle" scheme to a payment based on highway performance and availability. Finally, in all countries, the team found examples of the temporary transfer of existing government assets and revenue sources to the private sector. Transfers appeared in a variety of methods, from maintenance to tolls, for durations of up to 35 years.

CONCESSIONS

While the only a minimal number of quasi-public concession and private transportation projects have been developed in the United States, the European countries visited are leveraging concessions for major portions of their highway systems. Portugal, for example, has gone from 431 km of concessions in 1991, to a planned 2,700 km of concessions in 2006—representing 90 percent of its national highway network. The concession system is allowing Portugal to complete its strategic National Road Plan by 2006, an 8-year acceleration over the projected timeline of traditional methods. Concessions are used for both construction and maintenance of European motorways. Concession periods vary, but were commonly found to be 30 years. The Dutch are promoting concession periods that equal 75 percent of the design life of the product. Both public agencies and concession companies commonly obtain long-term warranties from their contractors, but the team observed widespread use of maintenance contracts in lieu of warranties. A variety of concession structures were observed, ranging from fully private to quasi-public and fully public entities, with varying requirements for private-sector equity. This report includes a discussion of a "Public-Private Comparator" employed by both the Netherlands and the United Kingdom in making procurement decisions. Drivers for the use of concessions range from lack of public funding to a belief that private financing and maintenance delivers a higher quality product and provides benchmarks for public-sector performance. Concessions also are discussed in the performance contracting section of this report.

CONCLUSIONS AND RECOMMENDATIONS

U.S. highways agencies should better utilize the efficiencies and resources that the private sector has to offer, through the use of innovative financing, alternative contracting techniques, design-build, concessions, performance contracting, and proactive asset management. Agencies must focus on the users, while equitably allocating risk and seeking to establish an atmosphere of trust in the implementation of procedures. This report presents a number of tools to assist U.S. agencies in meeting their growing infrastructure needs. Documentation of knowledge and best practices learned on the scan is provided in an effort to implement these tools and make the U.S. transportation system more efficient and effective for the public.

The team found a number of contract administration tools and techniques that will impact the U.S. transportation community. Some of these items can be directly and immediately applied, while others will require legislative changes prior to implementation. All team members will be actively taking opportunities to educate their peers about the results. Additionally, the following actions will be taken by the team to implement the most pertinent findings:

- Employ best-value techniques in the selection of construction professionals wherever it is shown that value can be added through quality or innovation.
- Explore techniques to fairly and equitably use confidential negotiations as part of the procurement process, as well as discussions of alternative designs and alternative bids to capitalize on the creativity and innovation of the private sector.
- Create specifications that define the owner's performance objectives, which can be used nationally to promote consistency in performance specifications while allowing for innovation in design, construction, and maintenance.
- In conjunction with the performance specification system, develop consistent and objective performance indicators that allow for the measurement and verifiable benchmarking of the performance specifications nationally. These performance indicators should be used to create a system of continuous improvement of outcomes for the industry.

The following table is provided as a guide to the report. Europeans are using certain tools to assist in solving their transportation needs. The table summarizes the tools discovered on the scanning tour and correlates these tools to the needs of U.S. highways agencies.

INFRASTRUCTURE NEEDS AND CONTRACT ADMINISTRATION TECHNIQUES

	Growing Infrastructure Needs	Inadequate Access to Public Funds	Insufficient & Diminishing Staff	Lack of Innovation in Delivery	Slow Delivery & Delays	Cost Overruns	Adversarial Relationships	Claims Oriented Environment	Perceived Lack of Maintenance Efficiency	User Frustration	Political Discontent
Chapter 3: Contracting Techniques											
• Procurement											
○ Best-value Selection											
○ Alternative Bids/Discussions											
• Contract Types											
• Payment Mechanisms											
Chapter 4: Design-Build											
Chapter 5: Performance Contracting											
• Performance Specifications											
• Performance Indicators											
Chapter 6: Alternative Financing Techniques											
• Funding Sources											
○ Public-Private Partnerships											
• Payment Mechanisms											
○ Shadow Tolls											
○ Active Management Payment Mechanism											
Chapter 7: Concessions											

Chapter 2: Introduction

BACKGROUND

Traditional methods of contract administration have remained virtually unchanged in the U.S. public highway industry for more than 50 years. The traditional system of contract administration involves public funding of highway projects in a "pay-as-you-go" manner in combination with a two-step process of procurement that clearly separates design from construction. Design, for the most part, has been done by the public highway agencies and construction has been procured from the private sector in a low-bid approach. The traditional method is time consuming because of the linear nature of the design and construction process. The separation of design and construction, in conjunction with the low-bid environment, has often led to a culture of claims and substantial cost increases. Although tested and familiar, the traditional method of construction administration in the U.S. highway industry is under increasing pressure to undergo changes to better meet increasing infrastructure needs.

In the past 10 to 15 years, public agencies have begun to employ a wider variety of contracting procedures as a result of increasing traffic, deteriorating infrastructure, and diminishing staff. The growing interest in alternatives is evidenced by the number of participants in the FHWA Special Experimental Project - No. 14, the Federal Transit Authority's turnkey program and in the American Road and Transportation Builder's Association (ARTBA) committees on public-private ventures. Even with the most generous estimates, however, only a very small percentage of current U.S. contracts fall into the alternative contracting category. Recognizing that European countries have significantly more experience in the use of alternative contracting procedures, a team of federal, State, private-sector and academic researchers was organized to observe and document those contract administration processes that might have value to the U.S. industry.

PURPOSE AND SCOPE

The purpose of the scan trip was to observe and document alternative contracting practices in Europe in order to transfer best practices and lessons learned to the U.S. highway industry. The American highway community has a high level of interest in improving contracting procedures and practices throughout the United States. The widespread experience gained by European countries offers the United States valuable insight into the problems and solutions associated with using these innovative techniques. Because of sufficient similarities between Europe and the United States, many concepts should be transferable.

In some sense, this scanning tour was a followup to the 1994 Contract Administration Techniques for Quality Enhancement Study Tour (CATQUEST) and to asphalt and concrete paving scans in the early 1990s. Because such a large number of the findings from the CATQUEST study are being employed in the United States, the FHWA/AASHTO/NCHRP consortium on International Programs saw value in exploring the subject further. The 2001 contract administration tour was formed in part to see what subsequent lessons the Europeans had learned on the topics identified in the 1994

CATQUEST visit and in part to seek new related topics that could impact the U.S. highway industry in a similar manner as those discovered 7 to 10 years ago.

The scope of study involved traveling to European host countries with the most activity in the areas of interest, to research and document best practices that might benefit U.S. practitioners. The specific areas of interest included innovative financing, alternative contracting techniques, design-build, concessions, performance contracting, and asset management. Upon returning to the United States, the team was charged with disseminating the results and implementing those findings with the greatest potential to improve the industry.

Given the limited time afforded to the study team during the visits to the host countries, it is important to note that this report does not include all of the contract administration techniques used in the countries visited. Instead, this report intentionally focuses on programs and techniques that the study team believes have the greatest potential benefit for further consideration and implementation in the United States.

METHODOLOGY

The contract administration scan was selected by the Transportation Research Board's (TRB) NCHRP's Panel 20-36 from a number of competing proposals for the 2001 funding cycle. Upon acceptance of the proposal, two co-chairs were named as representatives for the funding agencies: David Cox, Oregon Division Administrator for the FHWA, and Ron Williams, State Construction Engineer of the Arizona DOT for AASHTO. They in turn chose representatives from the public and private sectors to represent a cross-section of the industry, as follows:

David O. Cox (Co-Chair)
FHWA

James J. Ernzen
Arizona State University

Charlie Franklin (Frank) Gee
Virginia DOT

Gregory G. Henk (Representing ARTBA)
Flatiron Structures Company, LLC

Jeff W. Kolb
FHWA, California Division

Tanya C. Matthews, AIC
Design-Build Institute of America

Keith Molenaar, Ph.D. (Report Facilitator)
University of Colorado

Len Sanderson
North Carolina DOT

Nancy C. Smith
Nossaman, Guthner, Knox & Elliott, LLP

Gary C. Whited
Wisconsin DOT

Ronald C. Williams (Co-Chair)
State Construction Engineer for the Arizona DOT

John W. Wight (Representing ARTBA)
HNTB Corporation

Gerald Yakowenko
FHWA

The next step was to conduct a "desk scan" to select the most appropriate countries to visit. The objective of the study was to maximize the time spent by the panel in reviewing topics of interest. This desk scan employed a three-tier methodology of literature review, expert interviews, and synthesis. This methodology provided for data collection from government agencies, professional organizations, and experts abroad who are most advanced in the selected topics. The desk scan was very useful. For instance, none of the team members had suggested Portugal as a country to visit, nor had any of the previous scans visited Portugal. Because of its innovative and extensive concession program, funded in part by the European Investment Bank, Portugal was revealed to be one of the European countries most active in the contract administration topics of interest. For a copy of the 2001 Contract Administration Desk Scan, please contact the Office of International Programs with the FHWA (www.international.fhwa.dot.gov).

After the host countries were selected through the desk scan, the team finalized a "panel overview" document. The panel overview was sent ahead to the host countries to prepare them for the U.S. delegation. The panel overview explained the background of the study, the scope of the study, the sponsorship, team composition, topics of interest and the tentative itinerary.

Prior to conducting the scan tour, the team prepared a comprehensive list of amplifying questions to further define the panel overview and also sent those questions ahead to the host countries. The process of assembling the final list of questions took several iterations, with a final team meeting 8 months prior to the scanning tour. Some of the host countries responded to these questions in writing prior to the scanning tour while others used the questions to organize their presentations. An attempt was made to craft the questions precisely enough that the team would not miss any information that it anticipated, yet open-ended enough that new ideas—not envisioned by the U.S. scan team—could be brought to light by the host countries. The team was successful in its assembly of the questions, as seen by the answers to all of the questions and new topics added as documented throughout this report. Appendix C contains a copy of the amplifying questions that were sent to the host countries.

The delegation traveled to Europe from June 6-24, 2001. The visit consisted of meetings with highways agencies and practitioners as well as site visits. The scan team visited:

- Lisbon, Portugal
- The Hague, the Netherlands
- Paris, France
- London, England
- Kettering, England

PREVIEW OF THE REPORT

The report combines definitions and illustrative case study examples of contracting techniques in Europe with critical analysis of the applicability of these techniques to U.S. contracting. Whenever possible, U.S. parallel examples are provided to amplify those techniques that are directly applicable. The report is organized in the areas of alternative financing, contracting techniques, design-build, concessions, performance contracting, and asset management, as shown in the figure below.

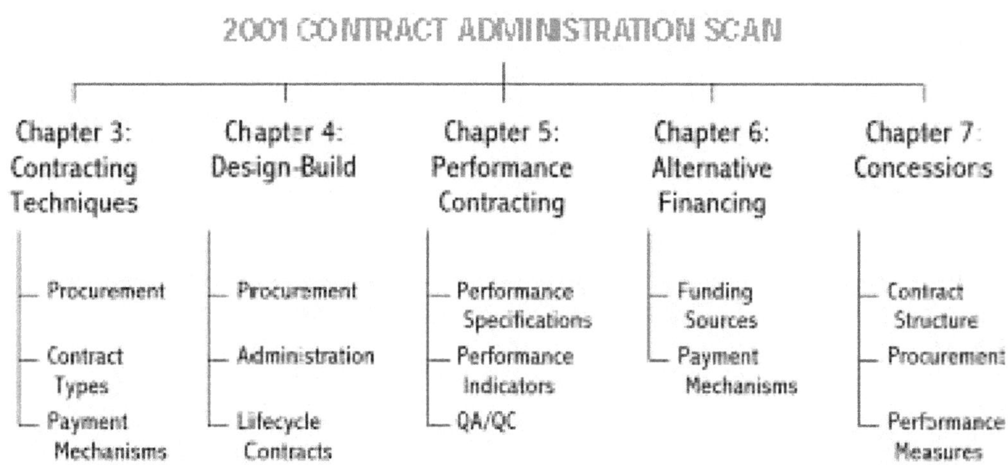

Chapter 3: Contracting Techniques

Alternative contracting techniques used by European transportation agencies could have a tremendous impact on the efficiency and effectiveness of contract administration in the United States The report discusses these techniques in terms of procurement, contract types, and payment mechanisms. Similar to the U.S. relationship between the State DOTs and the FHWA, the EU directives establish minimum requirements for procurement, but individual countries can develop unique contracting techniques that fit distinctive needs just as State DOTs do in the United States. Consequently, the contracting techniques documented in this report have the potential to be directly implemented in the United States.

The primary findings of this chapter involve the widespread use of *best-value procurement*, greater latitude to enter into *competitive negotiations*, more use of *alternative designs* in proposals, extensive use of *management contracts*, long-term contracts *tying maintenance to construction,* and *payment methods* that are based on *outcomes* at the end of the projects rather than payment for work as it is put in place. All of these techniques result in contractors and designers working more closely with the public sector and being given more public trust. Many of the techniques resemble contracting techniques employed in the U.S. private road sector and public building sector, but they have not for the most part been employed in the U.S. public transportation sector.

The United States and Europe share the same traditional method of contracting: the design-bid-build methodology. In the 19th century model developed in the United Kingdom and employed in many European countries, the client procures a professional design team separate from the contractor and awards construction contracts based on low bid. Professionals are employed to supervise the contractor. During the 1980s, a slowdown in development and consequent overcapacity meant that construction bids included low profit margins. This situation led to confrontational behavior, claims specialists, increasingly expensive claims submitted, years to settle, and unacceptable time and cost overruns for clients. An example provided by Yogesh Patel of the British Highways Agency included a project bid at 31 percent less than the engineer's estimate. Claims pending 1 year after construction would have brought compensation of 111 percent over the bid amount. The parties finally settled the claims for 42 percent of the bid amount—11 percent over the original engineer's estimate.

Likewise, in the Netherlands in the mid-1990s, the Dutch were experiencing concerns regarding the lack of innovation, efficiency, and competition. The result was a countermovement toward integrated solutions and project development methodologies such as design-build. This change involved transfer of technical risks to the private sector, giving the private sector greater influence on design and development of larger public works projects, and challenging the private sector to provide innovation both in product and process. The primary goal was to achieve more efficient solutions; a secondary goal was to improve opportunities for Dutch businesses to compete outside of the Netherlands.

Following is a quote from the Construction Industry Research and Information Association (CIRIA) report *Contract Incentivisation Schemes: Lessons from Experience* (Richmond-Coggan, D. 2001, p. 39), which summarizes this need for transition away from the traditional procurement methods. “Traditional procurement routes in the public sector may appear to be tortuous and leave little to the imagination in respect of expediting best-value.” The report goes on to suggest “a more open style of procurement against the backdrop of public accountability … The recent challenge to traditional procurement strategies within the public sector by the private finance initiative (PFI) has shown that the constraint on openness can be lifted to encourage the use of incentives within the negotiation process and their inclusion in the contract. Whilst this is an entirely different procurement methodology it has combined the issues of risk and reward within the negotiations and developed more openness in aligning the client and contractor objectives.”

All of these issues have influenced the Europeans to turn toward alternative contracting methods. Standard EU regulations (discussed below) and use of alternative contracting methodologies by many different EU members have validated the alternative methodologies, making it easier for EU members wanting to use alternative methodologies to obtain internal legislative approval to proceed. Innovations in procurement, contracting methods, and payment methods have resulted in an enhanced collaboration with the private sector. The European construction industry is beginning to understand the mutual benefit of long-term relations and managing its supply chain.

The European Parliament and the Council of the European Union have adopted a directive regarding rules to be followed by EU members in procurement of public works contracts and concessions over a specified amount. See Council Directive 93/37/EEC of June 14, 1993 (http://www.tendersdirect.co.uk/thelaw/ecd9337.asp), as amended by Council Directive 97/52/EC of October 13, 1997 (http://www.bipcontracts.com/directive7.htm). The Commission of the European Communities has proposed certain amendments to the Directive providing for electronic purchasing, expanding the ability to negotiate contracts, providing for framework contracts, clarifying provisions relating to technical specifications for the purpose of encouraging effective competition, strengthening provisions relating to award and selection criteria, simplifying thresholds, and providing a common procurement vocabulary. See http://europa.eu.int/comm/internal_market/en/publproc/general/com275en.pdf. Council directives are implemented by individual EU member states through appropriate governmental action. For example, the above-cited Directive was implemented into U.K. law by Public Works Contracts Regulations 1995. See http://www.bipcontracts.com/Briefings/Briefings2000/Brief7_00.htm.

The goal of the Directive is to ensure that contracts over a certain value are awarded in a competitive, transparent, and nondiscriminatory manner. The EU procurement requirements flow through to local agency contracts for which the state provides more than 50 percent of the funding. Furthermore, all contracts (including those under the specified threshold) are subject to certain requirements set forth in the treaty establishing the European Communities: no discrimination on the grounds of nationality; free movement of goods; a prohibition of quantitative restrictions on imports and exports; and measures having equivalent effect, freedom of establishment, and freedom to provide services. See http://www.ntc.no:8088/01/25/systm001.pdf for answers to frequently asked questions.

The EU Directive provides for three different types of procurement: Open Procedures allowing all interested contractors to submit tenders; Restricted Procedures whereby only those contractors invited by the contracting authority may submit tenders; and Negotiated Procedures whereby contracting authorities consult contractors of their choice and negotiate the terms of the contract with one or more of them. For negotiated procedures to be used, the agency must make certain findings. The proposed amendment to the Directive would expand the ability to use negotiations. Sole source negotiations are permitted only in limited circumstances.

Under both the existing and proposed Directives, award may be based on either lowest price or may be made "to the most economically advantageous tender." Under the proposed amended Directive, complex contracts would have to be awarded on the latter basis. For contracts awarded based on a determination of economic advantage, the contracting authority must state in the contract documents or in the contract notice all the criteria it intends to apply to the award (such as quality, price, technical merit, aesthetic and functional characteristics, environmental characteristics, running costs, cost-effectiveness, after-sales service and technical assistance, delivery date, delivery period, or period of completion). The original Directive stated that the criteria must be listed, "where possible," in descending order of importance. The amendment would require the relative weightings to be disclosed.

PROCUREMENT

Best-Value Selection

The most notable difference between European and U.S. procurement methods is that best value (referred to in the Directive as "most economically advantageous tender") is used in virtually all types of procurements. Best-value selection involves the evaluation of technical and management factors in addition to cost—as opposed to the low-bid selection, which involves only cost comparison of responsive bids from responsible bidders. Although the EU Directive permits low-bid selection,

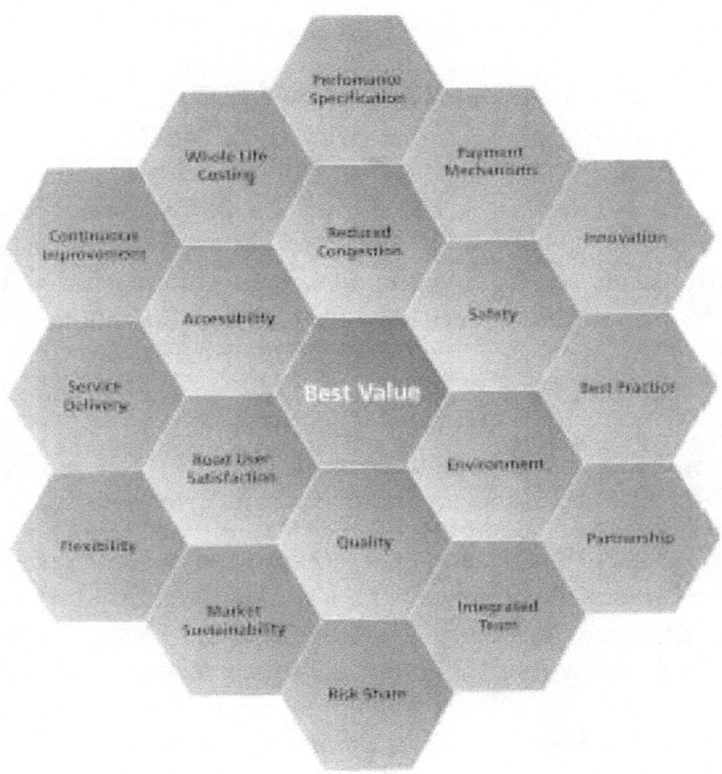

The British Highways Agency's goal of best-value selection.

in general it appears that low-bid selection is limited to relatively simple projects. The Europeans have found that best-value selection, using transparent and uniform processes, enhances competition and innovation. In cases of long-term maintenance contract procurements, the business culture and quality are weighted much more significantly than the price and technical portions of the procurement. There also is widespread use of prequalifications in selections to ensure technical competence in the procurements. The British Highways Agency's overall goal of best value is shown in the adjacent figure (British Highways Agency 1997).

Although all countries visited employed at least some type of best-value selection, the specific best-value criteria and weighting of these criteria varied from project to project and country to country. For example, Portugal includes best-value factors such as schedule and quality of technical proposal, but qualifications of proposers are reviewed on a pass-fail basis.

The Netherlands has a slightly different approach. For most construction projects, combinations of contractors (consortiums) compete for the contract. Depending on risks for the government, they use a process involving initial shortlisting followed by evaluation of final proposals. The shortlisting is based on evaluation of criteria establishing the contractor's capability of performing the contract, and can include competence, resources (experienced staff, special equipment, process certificates, etc.), experience and achievements, work quality in previous projects, approach to project (if required), and execution plan. Generally, the Netherlands applies the same evaluation criteria to review of the final proposal as for the shortlisting, plus price.

The typical Swedish best-value procurement is comparable to that used by the Netherlands. The Swedes typically employ a 70 percent price weighting with 30 percent weighting of references, schedule, Q/A system, traffic safety, environmental issues, etc. The Swedes typically do not publish the numerical weights of the evaluation criteria but instead rank the relative importance of each factor. Their legal counsel has determined that this approach meets EU requirements.

In U.K. design-build projects, the first contracts were awarded based on 20 percent quality, 80 percent price. Currently, weighting of 60 percent quality and 40 percent price is more standard, and sometimes quality is given an even higher weight. In some long-term maintenance contracts in the United Kingdom, a weighting of 90 percent has been placed on factors other than price.

The French have a best-value system that most closely resembles the low-bid system. They use an annual prequalification process for certain types of projects. Evaluation criteria are identified in the request for tenders in the order of priority (schedule, quality, price, etc.). Usually price is not the top criterion, but in practice the French Ministry of Transportation stated that the low bidder is selected in 95 percent of the cases. This situation occurs in part because technical proposals typically are not detailed enough to allow selection based only on technical qualifications. It generally takes 2 to 3 weeks to review the bids. Most of the situations involving selection of other than the low bidder are with alternative requests for proposals (RFPs). It usually takes 1-1/2 to 2 months to review proposals when the alternative process is used.

U.S. Parallel: Best-Value Selection

The United States is beginning to employ best-value selection, but primarily in design-build contracts where there is opportunity for innovation and minimal design is provided by the transportation agency. FHWA's draft design-build rules state that no less than 50 percent of the selection be based on price. In practice, price is most often the highest weighted factor. As in Europe, the methods for combining price and qualifications and the weighting of the technical factors vary from agency to agency. The NCHRP has completed one study of best-value procurement (NCHRP Report 451 – *Report Guidelines for Warranty, Multi-Parameter, and Best-Value Contracting 2001*) and, at the time of this report, has begun a second study to provide guidance for U.S. highways agencies in procedures for best-value selection.

Best-value selection has the potential to provide for the selection of higher quality contractors while lowering costs by rewarding technical innovations. Best-value procedures also have the potential to create a less adversarial and claims-oriented industry by developing a higher level of trust between the public and private sectors. Employment of a best-value selection system will also create accountability for past performance, as it can be incorporated into the selection criteria for future projects.

Confidential Discussions of Alternative Designs and Alternative Bids

Ministries of transportation visited use confidential discussions of alternative designs much more readily than in the United States They advised the scan team that this practice has resulted in an increase in design and construction innovation. The competitive proposal and negotiation process benefits the government since it requires the contractors to be both innovative and cost-conscious. The EU process is generally acceptable to contractors, thus promoting competition. Specifically, the minimum EU standards promote relative uniformity of the procurement process throughout the EU and make it easier for contractors to compete all around the EU. The negotiation process enables contractors to benefit from innovations they propose without concern that their ideas will be shared with competitors.

In the United States, the standard low-bid procurement process does not allow contractors to incorporate alternative concepts in their bids, although design-build procurements often include such opportunities. For traditional contracts, a bidder's innovative ideas would be submitted only as proposed changes after contract award through some form of value engineering process. There is often a shared savings clause in the contract to create an incentive for the contractors to submit cost-savings ideas. The scan team observed numerous examples of contractor-initiated changes being integrated into the procurement process. These changes can be referred to as "alternative proposals," with the innovation and competitive pricing occurring during the procurement process, not after award. This process leads to better pricing on alternative designs, but it also requires longer bid review times.

A primary barrier for alternative contractor proposals in the United States is our inability or unwillingness to hold confidential discussions with proposers during the procurement period. The U.S. process only allows discussions during a competitive bidding process to clarify existing plans. Typically, any information shared with one bidder must be shared with all of the bidders. Procuring agencies are not allowed to discuss possible innovative proposals with only one team. Because teams do not want to have their alternative concepts shared with other teams, they are unwilling to ask questions about alternative proposals until after the contract has been awarded. This process results in less value to the owner. Some of the European agencies visited allow confidential discussions of innovative ideas during the procurement process and have subsequently experienced increased innovation and more competitive pricing. The new high-speed rail link from Amsterdam to the border of Belgium, discussed in the following section, provides an excellent example of confidential discussions during the procurement period for a design-build project.

The Netherlands High-Speed Line South

High-speed rail is linking the Netherlands to other countries in the EU. The new High-Speed Line South is running between the Skidmore airport and Rotterdam. Part of the line runs through the Green Heart of the Netherlands, an area designated by the government for noncommercial use. This area is so important to the public that the government decided to run the high-speed rail in a bored tunnel under the area rather than disrupt its pristine nature during and after construction.

The bore tunnel selection included a prequalification phase using a Request for Qualifications (RFQ). Eight contractors responded to the RFQ, and the government shortlisted the teams down to five contractors to develop designs and bids. The bidding phase involved reviewing 10 bids, 2 proposals from each of the 5 bidders. Three of these 10 proposals were then chosen for competitive negotiations.

The final winning proposer's design had very little resemblance to the government's original design. The original design involved a shorter, two-bore tunnel with connections between the bores. The winning firm proposed an extremely large, one-bore tunnel that was actually longer than what the original design required. The result was a longer and safer tunnel that met the needs of the owner better than the original design.

The key to the success of this procurement was that the government agency was able to entertain design alternatives during construction that it did not share with all other bidders. The winning proposer had the ability to find out whether the government would entertain the idea of a one-bore tunnel, before spending the money to explore the idea with the tunnel boring machine vendor. The proposer had confidence that this information would not be distributed to other teams during the procurement process, thereby negating their competitive advantage. Note, however, that the government only commented on the feasibility of the alternative design solution during the procurement. It did not actually approve the new design until all bids were submitted.

U.S. Parallel

Various U.S. agencies have specifically asked proposers for alternative technical concepts, usually in the context of design-build projects. Such an approach was used for the Transportation Expansion (T-Rex) project, a multimodal project jointly undertaken by the Colorado DOT and the Regional Transportation District that included 19.1 miles of new double-tracked light rail transit and 16.6 miles of highway improvements. The prequalified teams submitted a total of 44 alternative configuration concepts for the T-Rex project. The agencies accepted approximately half of the proposals, and conducted three 6-hour workshops with each team to discuss the concepts. Although no formal analysis has been conducted regarding the benefits of this process, the agencies believe that the alternatives resulted in significant savings of both time and money.

CONTRACT TYPES

The scan team observed numerous contract types and delivery structures on the tour. The United States is currently employing a number of these techniques, but the scan revealed new techniques that have merit for implementation in the United States. This section specifically discusses those contracts that are not currently being used in the United States. Design-build, design-build variations, and concessions are introduced below, but are discussed in more depth in Chapter 4: Design-Build and Chapter 7: Concessions.

Contracts Similar to U.S. Methods

- Design-Build
- Design-Build-Maintain
- Design-Build-Operate-Maintain
- Concessions

Contracts not Currently Used in U.S.

- Framework Contracts
- Management Agency Contracting (MAC)
- Private Finance MAC
- Integrated Supply Chain Management

All of the contracts listed above promote methods of creating more partnership between the public and private sectors. European contracts have evolved toward placing more public trust and responsibility in the private sector. The contract types discussed in this report provide examples of how some European countries are reallocating contractual risk to leverage the efficiency of the private sector.

Design-Build

The design-build contract involves one contract for both the design and construction. The traditional method of contracting separates design and construction to create a system of checks and balances for quality and price. Although this separation creates checks and balances, it also can create a long delivery period and may result in an overdesigned project and an adversarial and claims-oriented environment. Design-build contracts speed the delivery of projects and promote more constructability and innovation.

European Highways Agencies have used design-build contracts to a much greater extent than U.S. agencies. Design-build requires a higher level of trust and cooperation between the public sector and industry than does the traditional method. As mentioned throughout this report, the public European highways community has a much closer relationship to the private sector than does the U.S. Specific lessons learned about design-build in Europe are discussed in Chapter 4: Design-Build.

Design-Build-Maintain and Design-Build-Operate-Maintain

The extension of a design-build contract to a design-build-maintain (DBM) or a design-build-operate-maintain (DBOM) contract is becoming prevalent in Europe. Adding maintenance or operation and maintenance has numerous advantages. The primary advantage is that these contracts create a lifecycle responsibility for the design-builder. The same company that designs and constructs the highway is also responsible for maintaining quality over a period of years. This situation provides an incentive to deliver better quality in the initial design and construction of the project because the design-builder will be responsible for additional maintenance and repair costs if the initial quality is inadequate. DBM and DBOM contracts are discussed in more detail in Chapter 4: Design-Build.

Concessions

The French national highway system is almost exclusively operated by concessionaires. Interestingly, only one of France's concessionaires is privately held; the remainder are limited liability companies owned by central and regional government bodies. Portugal is the most aggressive employer of concessions. Portugal has integrated the use of concessions into its long-term planning and anticipates expansion of its concession contracts from 431 km in 1991 to 2,269 km in 2006. Although the United States has limited experience with real toll concessions, European agencies have significantly more experience with other variations of concessions. The intricacies and implications of concessions are discussed in Chapter 7: Concessions.

Framework Contracts

The framework program was implemented in the United Kingdom in 1999. The framework contract is an arrangement that allows a purchaser to package its procurement requirements and select one or several suppliers to meet specific task(s) or order(s) over a period of time. The purchaser and suppliers establish terms on which purchases will be made at the outset, but do not set precise quantities. Frameworks can be applied to supply, works, and professional service activities, but they are best suited to orders of a similar nature, where demand is likely to materialize in a regular or programmed manner over an extended period. Frameworks enable purchasers to place orders, or "call-off" services, with or without secondary competition—substantially speeding procurement. As noted above, the proposed new EU Directive specifically permits such agreements.

Carillion, a framework contractor in Kettering, England, hosted the scan team. Carillion is responsible for the long-term maintenance of a section of highway on the national motorway. Carillion performs all maintenance (landscape, stripping, inspections, etc) on the highway and is the sole source for improvements on small construction projects. Carillion has a 5-year contract to perform maintenance and small construction contracts; it also can compete in the open marketplace on larger construction projects. Some of this work is self-performed by Carillion and some is done through its supply chain.

The following is a list of essential characteristics of a framework agreement as specified in the "Procurement Guidance Strategy Note" of the British Highways Agency's (HA) Works Framework Contract, Issue 2, Revision 0.

- The Terms and Conditions contained within the new model contract forms have a number of differences from contracts used for single procurement actions. They provide for a longer-term and closer relationship between the

Highways Agency and its supplier than that traditionally employed. This is to allow a process of continuous improvement to bear fruit. Typically works frameworks should be a minimum of 5 years duration. [Note, however, that the proposed EU Directive would limit the duration to a maximum of 3 years except under exceptional circumstances where a y-year [[**check**]]term is justified.]

- To achieve best-value through the use of a framework great clarity of its purpose is required. This includes clarity of the specification, the type of supplier required as well as clarity of the documentation. That is, five years is a long time if things are going wrong or one has the "wrong" supplier. The need to demonstrate continuous improvement, and the mechanism to deliver it, must be clearly defined.
- Unlike single procurement actions the quantum, timing and logistics for each procurement action planned under the framework is not likely to be known at the outset. Yet price competition remains important. A balance must be struck between burdensome pricing requirements and having sufficient price data to allow fair and competitive prices to be paid during the life of the framework. The importance of adopting consistent approaches to frameworks (i.e., the models) and use of benchmarking processes on performance and costs should not be overlooked. It is a key element of the control process and a main driver for continuous improvement.
- In addition the fundamental payment mechanism for a framework should be considered carefully. In pursuing best-value, requirements for price certainty may be relaxed within a framework contract, as there is the incentive of a continuous income stream if the supplier demonstrates quality and value, i.e., if they do this job well there will be another to follow. There are many payment mechanisms that can be used for frameworks but in deciding which one to use a risk assessment should be made and the method that is likely to attract the smallest risk premium from the supplier should be selected.
- For a starting point, the HA has decided to adopt a target cost payment mechanism for its Model Framework works contract as this offers control with cost openness and an opportunity to deliver continuous improvement.
- Frameworks should not be entered into lightly. To get value out of them the Highways Agency must be offering a consistent workload of similar activity for a medium to long-term. Care should also be taken that we do not overstate what we can deliver. There must be sufficient demand for a framework and that demand must extend over a number of years for value to be achieved. Much of the value for money will come from the efficiencies of the process and the efficiency gains that repetition brings.
- In order for the HA to deliver the required consistency of workflow it may be necessary to have a framework contract extend over a region (not just an area) or even nationally. At the same time the impact on the supplier market must be considered in order to maintain competition and a commercially sustainable market into the future.

- The mechanism for placing orders must be considered carefully. If one supplier is used control is straightforward. However, with multiple suppliers one person needs to manage the framework as a whole and the mechanism for distribution of work amongst the suppliers should be established and explained to them in advance of the tender process. The HA Model Framework Operational Guidance Notes include a suitable method and suggests that a Framework Board be established to manage the process.

Framework contracts have been found to offer flexibility, speed of delivery, quality and reliability of supply or service, and value for money. However, agencies must be able to demonstrate that best value is being achieved at the outset and throughout the long relationship with the supplier. This outcome can only be achieved through demonstration of continuous improvement.

U.S. Parallel – Indefinite Delivery/Indefinite Quantity Contract

A U.S. parallel to the framework contract is the Indefinite Delivery/Indefinite Quantity (ID/IQ) contract used by the federal government. ID/IQ contracts have been used for some time by the Department of Defense for the procurement of military equipment and have recently been employed for design and construction of capital facilities. A small number of contractors (typically three to five) are placed under contract for a 1- to 5-year term to deliver equipment such as artillery or vehicles. An indefinite-quantity contract provides for an indefinite quantity, within stated limits, of supplies or services during a fixed period. The government places orders for individual requirements. Quantity limits may be stated as number of units or as dollar values. Recently, government agencies have begun to procure design and construction services more frequently under ID/IQ contracts. Renovations of aging military bases and expansions where the quantity of work is not well defined have shown particular value. ID/IQ contracts may be very well suited for much of the routine maintenance on U.S. highways where the description of work can be clearly defined, but the quantity of work may vary depending on use, weather, or other circumstances. This type of work may include stripping, resurfacing, etc. For more information on ID/IQ contracts, consult the Federal Acquisitions Regulations (FAR) Subpart 16.5- Indefinite-Delivery Contracts.

Managing Agent Contracts and Private Finance Managing Agent Contracts

The managing agent contract (MAC) was first employed in the United Kingdom in 1996, and 24 contracts were in place in 1999. The managing agent is responsible for carrying out all design work, asset inspections, network maintenance management, and supervision of the term maintenance contractors. The term maintenance contractors are responsible for all routine, cyclical, and winter maintenance; and small capital maintenance and improvement works.

The Highways Agency's role in this type of arrangement does not resemble that of a traditional U.S. State Highway Agency (SHA), but that of a network operator that strategically manages the network. Their role is to procure services, not to provide them. As discussed in the previous section on best-value selection, they are utilizing quality criteria and not just price. The Highways Agency defines the performance outcomes that it is seeking and then audits and regulates the MACs to ensure quality.

The private financing managing agent contract (PFMAC) is an extension of the MAC concept in which the managing agent also provides financing for the capital and cyclical cost of the MAC agreement. The PFMAC is then paid for its services through a number of performance-based payment systems. MACs and PFMACs are discussed in greater detail in Chapter 5: Performance Contracting.

Integrated Supply Chain Management

A developing new contracting technique in Europe is integrated supply chain (ISC) management. This contract extends the use of managing agent and framework contracts to incorporate long-term strategic alliances with the suppliers of material and labor. The motivation for using ISC contracts is that approximately 80 percent of the cost of any manufactured product is in the supplier's labor and materials. Therefore, key suppliers should be selected for their ability to deliver excellent work at a competitive cost. The supply chain must be capable of contributing new ideas, products, and processes. The suppliers also should be managed so that any waste and inefficiency are continuously identified and driven out. Models of traditional, single source, and ISC delivery methods are shown below.

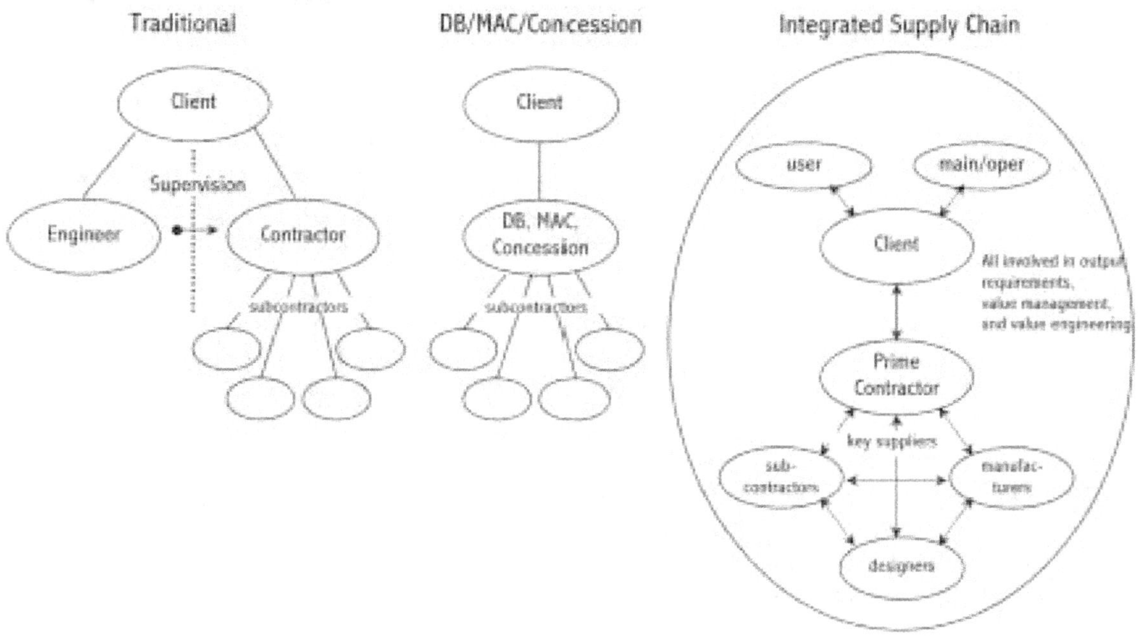

Traditional, single source and integrated supply chain contract models (adapted from CIRIA presentation).

The ISC methodology extends the sole source concept of the design-build, MAC, or concession contract by forming a strategic partnership with the prime contractor to manage key suppliers. The ISC concept leverages the fact that 80 percent of project costs are generated through materials and labor. Long-term contracts are formed with the key suppliers who can in turn become involved with the projects at their earliest inception. The designers, manufacturers, and subcontractors are involved in writing performance requirements, value management, and value engineering. The key suppliers also are much more involved in defining the needs of the users, operators, and maintenance personnel.

ISC contracts are new to the design and construction industry, but they are actually modeled after best practices in the manufacturing, automotive, and industrial sectors. Two examples of these best practices come from Shell U.K. and Rolls Royce. After oil prices collapsed in 1986 and again in 1992, Shell U.K.'s North Sea oilrigs were producing oil at a higher cost than it could be sold. The traditional rig design involved complex specifications that were developed and imposed by the operator with no reference to the supply chain. The specifications were requiring nonstandard materials and processes with inequitable contracts. The solution was to move to an output specification (no detailed specifications) and allow suppliers to propose optimal solutions. Shell U.K. and the suppliers developed the projects through mutual interest using nonadversarial techniques. They did this by taking the most economical bid, incentivizing the contracts, and capping the supplier's risk. Shell U.K.'s result was a 30 percent reduction in cost realized in 3 years.

Rolls Royce achieved a similar 30 percent reduction in the cost of producing three engine test cells that were critical to remain competitive. One of the keys to achieving this reduction was to abandon the competitive tender process and replace it with a target-pricing concept for the supplier. In essence, the target price was agreed upon early in the project and Rolls Royce shared the pain and gain of achieving the target price with the suppliers. Using an ISC concept in lieu of the traditional competitive tendering process led to savings in in-house costs, optimal designs, and removal of uncertainty of final price. Target pricing is discussed in the next section.

Although ISC contracts are not yet in full-scale practice in Europe, the United Kingdom is moving in that direction. The term maintenance contractor agreement in the U.K.'s MAC and PFMAC contract uses many ISC attributes. These contracts include long-term maintenance agreements with maintenance operators. The Highways Agency takes the role of a network operator and not a provider of services. The term maintenance contractor agreement aspect of the MAC is discussed in more detail in Chapter 5: Performance Contracting.

PAYMENT MECHANISMS

Alternative procurement methods and contracts require nontraditional payment mechanisms to optimize their benefits. The biggest difference from the traditional payment methods is that payments are not based on units of work completed, but on availability of the product at the end of the project. The private-sector providers are required to carry much more of the costs during and after construction. They are then paid on the basis of milestones, availability (i.e., number of lanes open), quality of performance (i.e., smoothness), and/or safety (a reduction in the number of crashes, measured against a baseline). In concession contracts, payments are evolving toward a purely performance-based method.

Since the majority of contracting techniques found on this scan and documented in this report involve design-build or concessions, there is little discussion regarding the traditional unit price contracts. The host countries did not even discuss the variations of unit price contract, such as cost-plus-time bidding and lane rental concepts invented in the United Kingdom in the late 1980s. These techniques have taken a smaller role between the agencies and the design-builders and concessionaires.

However, design-builders and concessionaire may use unit price or work order subcontracts depending on the scope of work for the subcontract.

At the government levels in France, Portugal and the United Kingdom, there has been a major shift away from provider of services to network operations manager, and the payment methods reflect this shift. The Netherlands still maintains a much more traditional role and, much like the United States, has only begun to experiment with the operator role. For traditional contracts, Rijkswaterstaat (RWS; the national government department in charge of infrastructure, including roads) uses units of work, however, their standardized units of work may be very detailed. They provide separate payments for delivery, compaction, and curing of materials and even adjustments for terrain. Product-oriented contracts may provide for similar line items; however, the pay items are focused on the individual products.

Milestone Progress Payments

Alternative contracting techniques often require a contract award with less than 100 percent complete designs. This fact makes unit price payment information unavailable at the time of award. Numerous European agencies have turned to milestone payments to overcome this problem. Milestone payments involve larger payments based on the completion of certain major work items. This method has been used successfully in the U.S. building industry for years. The primary barrier to milestone-payment methods in the highway sector is that it requires the contractor to carry a large financial burden in between milestone payments. Since the European Highways Agencies are using more private financing, this is not a significant barrier.

On design-build projects, the British Highways Agency will generally monitor the design-builder's performance to verify that certain milestones have been achieved, or audit its records if payment is based on invoiced costs. On design-build-maintain contracts in the Netherlands, payment may be based on the completion of certain construction stages or milestones. During the maintenance phase, RWS makes a fixed lump-sum payment every 3 to 6 months if the desired performance criteria are achieved. This payment method also is used on performance contracts.

The Netherlands also is paying by percent-complete formula on the Westerschelde Tunnel profiled in Chapter 6: Alternative Financing, and is very pleased with this payment procedure. Payments are made based on a percentage of progress. The contractor provides the owner with a percentage-based determination of completion every 2 weeks. The owner spends a minimum amount of time verifying progress; however, a detailed audit of the worked performed is done four times per year. RWS pays the contractor within 14 days of invoice. It should be noted that the project controls for this project are using the latest information technology to achieve the short payment time and to organize the audit structure.

Payment by Availability

Payment by availability means that the owner pays for a highway on the basis of the amount of time that the highway is open to traffic. This method is a radical change from the traditional payment for completed work by unit costs. The contractor, design-builder, or concessionaire is responsible for carrying all of the costs during construction and operation and gets paid only if the lanes stay open to traffic over an agreed upon period of time. The goal of this payment method is to reward the private sector for decreasing congestion and make it responsible for the entire project lifecycle from design through construction and maintenance.

Of the five European countries visited, the British Highways Agency is the most advanced in this payment method. As noted previously, the British agency has been going through an evolutionary process over the past 10 years. Although the cost-plus-time bidding and lane rental concepts were invented in the United Kingdom in the late 1980s, these techniques are no longer used. The agency also went through an evaluation period for design-build and design-build-finance-operate (DBFO) utilizing shadow tolls. The administrators at the agency now believe that the future contract forms that will be most economical will include payment by availability.

Over the past few years, the shadow toll contracting method has fallen out of favor with some administrators at the British Highways Agency. Opponents of shadow tolls claim that this practice actually increases traffic growth and is contrary to the overall goals of the transportation plan. For this reason, the agency has developed a modified contract and payment method for future DBFO contracts called active management payment mechanism (AMPM). Both shadow tolls and AMPM methods are discussed in Chapter 6: Alternative Financing.

The agency intends to rely on the congestion management provision in the AMPM method as a replacement for the shadow tolls concept. Under this new concept, the contractor's payment is based on the availability of lanes during the construction and operational period. A weekly payment will be made that is proportional to the expected level of traffic flow. This traffic flow will be based on target speeds and expected traffic volumes (in passenger car units/hour as a percentage of capacity). Adjustments will be provided for different time periods of the day and different geometric sections. In theory, this provision will provide the DBFO firm with a large contractual incentive to schedule construction and maintenance operations to minimize inconvenience to the traveling public. In addition, the DBFO firm will have an incentive to manage the traffic operations and improve the flow of traffic through the work zone. Such an incentive may encourage the firm to actively manage the work area by providing additional incident-clearing facilities, emergency towing vehicles, emergency pull-off areas, and other techniques to facilitate the flow of traffic.

The Netherlands is attempting a similar payment method. On the N31 design-build-maintain-finance contract, the progress payments are based on the availability of newly completed roadway sections. Even though the project is privately financed, RWS made the decision to make partial payments when roadway sections are open to traffic instead of waiting until the entire project is open to traffic. This system is intended to incentivize the builder to make the road available sooner, but share the risk for the entire project equitably.

Target Pricing

All of the European countries visited use some form of target pricing with incentives and disincentives. This concept is not unlike the U.S. construction-manager-at-risk concept, and also is the approach commonly used for private-sector design-build in the

United States. During the design development process, the owner establishes a target price, and the contractor will produce a design based on the project criteria, with reference to the target price. A guaranteed maximum price will be negotiated with the contractor when the design reaches a point where risks can be assessed accurately. Cost-sharing provisions, like value engineering provisions found in current U.S. contracts, are used to incentivize contractors to optimize design and construction costs. A model of target pricing is provided in the figure below.

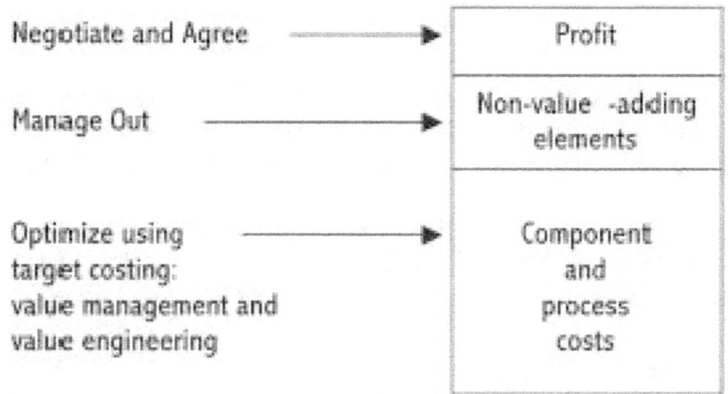

Target-pricing model (adapted from CIRIA presentation).

The target-pricing model has been employed in the U.S. building market for years, and good contract examples are available from the American Institute of Architects, the Associated General Contractors Association, the U.S. General Services Administration, and many others. The fundamental premise lies in setting an early target cost from engineering estimates, then capping the owner's risk through a guaranteed maximum price at some percentage above the estimate and incentivizing contractors to minimize costs by offering shared incentives for savings below the target price. The major barrier that exists in the U.S. highway sector is that the target-pricing system requires reliable conceptual cost estimates early in the project and an open book accounting structure throughout the project lifecycle. These two significant hurdles in the U.S. transportation market are less burdensome in Europe given the closer relationship between European agencies and the private sector (utilizing open book systems in many instances). Open records laws may present an additional barrier in the United States. The private sector guards its secrets carefully from competitors, and may believe that information shared with public agencies will somehow become public.

Incentives and Disincentives

Incentive and disincentive methods were embedded into many of the projects involving long-term maintenance structures that were presented to the scan team. The transference of finance, safety, and maintenance responsibilities to the private sector results in intensive auditing from the owner to ensure peak performance. These audits, in turn, can provide measurable metrics that translate into incentive and disincentive schemes. Two primary examples of incentives and disincentives come from the incentive plan for safety and quality management in the British Highway

Authority's AMPM contract discussed above and the penalty plan for The Netherlands quality system being used on long-term maintenance methods.

The safety management payment feature provides a contractual incentive if the concessionaire achieves a better than average safety record in comparison with accident rates on similar roads. The service management provision provides a contractual incentive (1 to 2 percent of the DBFO firm's annual revenue) if certain quality goals identified in the firm's management plan are attained. The safety and quality management methods for the AMPM system are discussed in more detail in Chapter 5: Performance Contracting.

In the Netherlands, bonuses are sometimes used to reward contractors for above-average work if no deductions were made for work below a certain quality standard. Although direct contractual incentives for quality and contract time are not provided, the RWS utilizes other contractual measures to ensure quality in construction and a minimum of construction time. The RWS uses a yellow-card/red-card system to control design, construction, and operations quality during the life of the contract. Similar to the yellow-card/red-card system in the game of soccer, the contractor's performance is monitored on a program basis to ensure that it is operating in accordance with its approved QA/QC plan. If a significant deviation is detected, the RWS will issue a yellow card to warn the concessionaire of substandard performance. A second yellow card may follow. Payment reductions may be associated with each yellow card on the basis of the issue involved. In a serious situation, a red card may be issued, which would result in cessation of the work and, ultimately, termination of the contract and possibly a removal of the firm's prequalification rating.

It also should be noted that a significant early completion incentive is inherent in the toll concession concept. Until new construction is open to traffic, the concessionaire will not receive a payment in the form of real tolls or shadow toll payments from the owner. Again, the contract structure incentivizes the concessionaire to consider the whole lifecycle of the highway, which will lead to a competitive balancing of lifecycle maintenance costs and early completion gains.

SUMMARY

The European highway community employs a wide range of contracting techniques. All of the alternative contracting techniques described in this chapter require a higher level of trust and teamwork between U.S. highway agencies and contractors than currently exists. The level of partnership witnessed in Europe came through slow change from the traditional contracting systems, and all of the problems have not been solved. If the United States wishes to benefit from the efficiencies observed in European highway contracting, it must be willing to take strides toward alternative contracting methods and transition from established contracting techniques to more open and transparent methods.

As discussed in this chapter, the European highway community is benefiting from widespread use of best-value procurement, greater latitude to enter into competitive negotiations, more use of alternative designs in proposals, extensive use of management contracts, long-term contracts tying maintenance to construction, and payment methods that are based on outcomes at the end of the projects rather than payment for work as it is put in place. The U.S. highway agencies can directly apply these findings to many of their existing contracting schemes. The scan team recommends that the following tools be explored in the United States as a means to speed the delivery of our infrastructure and increase the quality of construction and maintenance:

- Use best-value award techniques in the selection wherever it is shown that value can be added through quality or innovation.
- Explore techniques to fairly and equitably employ confidential negotiations and discussions of alternative proposals to capitalize on the creativity and innovation of the private sector.
- Use management contracting on repetitive work where project characteristics display a potential to save construction and procurement costs.
- Explore integrated supply chain contracts to capitalize on the efficiencies documented in the manufacturing sectors.
- Test payments by milestones and payments by availability as a way to tie quality performance to payment structures.

Chapter 4: Design-Build

Design-build delivery has been steadily increasing in the U.S. public building sector for more than 10 years, but it is still termed experimental in transportation. To date, under Special Experimental Project 14 (SEP-14) the FHWA has approved the use of design-build in more than 150 projects, representing just over half of the States. The European countries visited have used design-build delivery for a longer time than the United States and provided the scan team with many valuable insights. The primary lessons learned on this scan tour relate to the types of projects utilizing design-build, the use of best-value selection, percentage of design in the solicitation, design and construction administration, third-party risks, the use of warranties, and the addition of maintenance and operation to design-build contracts.

Design-build is a project delivery method that encompasses both project design and construction under one contract. One firm, or team, is responsible for a project in its entirety. Design-build contracts can have many different forms, but the key element is a single source of responsibility for both design and construction under one contract. There are numerous reasons why owners choose to use design-build, but the primary reason is the potential for shortened project duration. Because of coordinated efforts between the designers and the builders, construction can begin prior to the completion of construction documents.

A design-build contract allows the public-sector owner to shift risks to the private sector that have traditionally been assumed by the public sector. The primary risk that is shifted is that of design errors and omissions. Because the design-builder owns the details of the design, it is responsible for all errors and omissions that occur after award of the project. Conversely, for accepting this risk, the design-builder also is entitled to value engineering savings that are traditionally shared with the owner. When design-build is used in conjunction with financing or operation, the risk shift is much more significant.

The transition from the traditional design-bid-build process to design-build can be difficult, and nowhere is this more apparent than in the U.S. highway sector. Perhaps the two biggest obstacles for design-build in the highway sector are: (1) the customary (statutory in most States) use of low-bid contracting in public highway construction; and (2) the concern on the part of the government agency that it should have full responsibility for design approval and construction administration.

U.S. highway projects have almost exclusively been awarded via unit price bids using 100 percent complete construction drawings. Low-bid selection is difficult in design-build because design-builders must be hired before design is complete to take full advantage of time and cost savings. Hiring the design-builder early in the process allows for fast-track construction and more constructible designs. The low-bid mentality is deeply entrenched into both the States' and contractors' business operations. The Europeans have fewer problems with the low-bid obstacle because best-value selection procedures are ingrained into their business culture.

In the United States, the responsibility for design traditionally rests on the States' DOT or a design consultant who is an extension of the DOT. When a DOT hires a design consultant, there has traditionally been a rigorous approval of design. The trend in design-build is that DOTs no longer approve the design, but rather stipulate a set of design standards. At various points, the DOTs may conduct a review of the design, but the final accountability lies with the design-builder. In some U.S. States, sovereign immunity for the design defects is tied to approval of the design by a public employee. This is apparently not an issue in Europe. Design approval and construction administration also are less of a problem in Europe, where they more readily utilize contractor-controlled QA/QC programs and tie maintenance and operation into the design-build contract. In those countries where the European agency takes a network operator role, design-build is prevalent, but where agencies take the role of "providing" engineering services, design-build is used less frequently.

This chapter discusses design-build techniques used in Europe. Specifically, types of projects utilizing design-build, design-builder selection, percentage of design in the solicitation, design and construction administration, third-party risks, use of warranties, and design-build-operate-maintain are presented. Where concession and public-private partnerships were studied, design-build was inherent in the process. Public-private partnerships are discussed in Chapter 6: Alternative Financing, and concessions are discussed in Chapter 7: Concessions.

TYPES OF PROJECTS UTILIZING DESIGN-BUILD

Evidence of design-build use was found on all types all projects, ranging from green-field construction to pure maintenance contracts of existing roads. Design-bid-build, however, is still a fundamental project delivery tool as well. In the United Kingdom, the Highways Agency's contracting method of choice is design-build, and it has almost completely replaced the design-bid-build method. In Sweden, design-bid-build is the primary delivery method, but design-build has been the standard for bridge design and construction for the past 10 years. Design-build is still in its early stages in the Netherlands. Most of the Netherlands' design-build contracts concern technical fields in which RWS has little experience, such as drilled tunnels and special electro/ mechanic devices. In both France and Portugal, concessions are the primary delivery method for major highway projects. Design-build is inherent in the concession process.

The projects that were found to be best suited for design-build involved an opportunity for innovation. Innovation can include technical innovations or time-saving innovations. Conversely, in the Netherlands, design-build contracts were not considered to be appropriate in most roadway projects because of constraints on innovation because of the process for approving the alignment, which allows only ~1-1/2 meters of leeway on either side. Design-build is considered a good delivery method for specific structures and unusual projects in which innovation is desired or a combination of government and private sector know-how is essential.

In both Europe and the United States, finding projects with physical characteristics that lend themselves to design-build is important, but perhaps the most important characteristic determining design-build use was found within the owner organization. Owners who take the traditional role of providing design with their own staff may be less likely to see benefits from design-build. Design-build requires a release of control for design details and quality control. For those owners who desire a higher involvement in design details and design preferences, design-build creates a difficult relationship. Likewise, those agencies who are comfortable with quality control through quality audits seem to be more likely to use, and benefit from, design-build.

DESIGN-BUILDER SELECTION

Design-build contracts can use many procurement methods, as shown in the figure below. While traditional design-bid-build delivery almost exclusively utilizes low-bid procurement, low-bid procurement is not well suited for most design-build projects. Design-build contracts are typically awarded with less than 30 percent complete designs. Awarding a project prior to final design speeds the overall delivery process and promotes construction innovation during design, but it is not conducive to low-bid award because the scope is not fully defined. Although the private sector can turn to sole-source selection and negotiate with the preferred design-builder, the public sector typically does not have that luxury. Most European agencies discussed some type of best-value award in conjunction with design-builder selection.

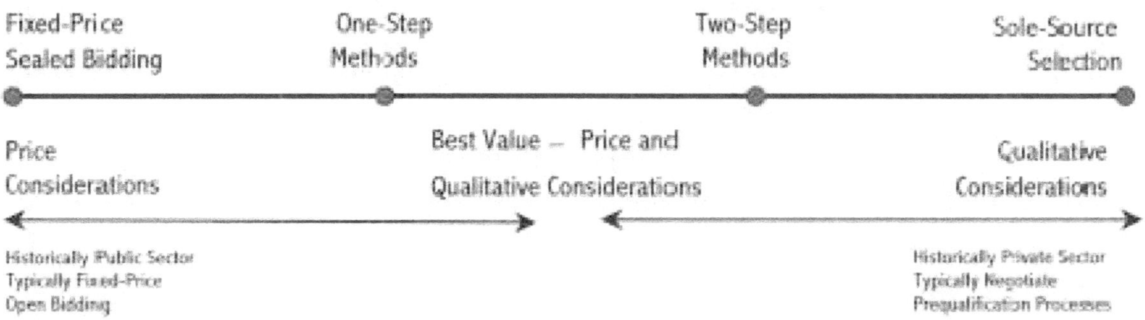

Design-build procurement continuum.

Best-value award procedures combine price selection with qualitative or technical factors in the final selection. It should be noted that most design-build projects discussed during the scan trip utilized a two-step process of prequalification followed by best-value selection. The first step serves to minimize the cost and effort of preparing a design-build proposal by the private sector and ensure that only the most competitive proposers move to the next step. In the Netherlands, as in all EU countries, the most important principle is and has always been free competition among all bidders fit for the job. Government tendering procedures also should be open, transparent, and objective. Depending on risks, RWS uses preselection and final contract allowance criteria. First-phase criteria are only tools to determine the fitness of a bidder in an objective way. In the Netherlands, these criteria can be:

- Competence: experienced staff, special equipment, process certificates, etc.
- Experience and Achievements: the quality of the proposers' work in past projects.
- Project Plan: proposers are sometimes obliged to deliver an execution plan.
- Preliminary Design: in more complicated design-build and public-private partnership (PPP) projects

The second step of the best-value selection involves the combination of price with other technical and managerial factors. This step typically is done in a linearly additive combination. For example, price bids are converted to points and added to the qualitative and technical points. The design-builder with the highest number of points is awarded the project—not necessarily the contractor with the lowest price. In U.K. design-build projects, the first contracts were awarded based on 20 percent quality, 80 percent price. Currently, weighting of 60 percent quality and 40 percent price is more standard; sometimes, quality is given even higher weight. Sweden utilizes varying cost and technical combinations, but the hosts estimated that the spread is typically 70 percent cost and 30 percent technical factors.

The final method of design-build selection witnessed on the scan was through the use of strategic partnering. Some countries are moving toward supply chain management type contracts. These contracts require long-term partnerships with selected design-builders and vendors. These selections are based almost exclusively on qualifications and past performance, as the scope of work is too nebulous to price accurately. The key to supply chain management is the creation of long-term relationships where the owners and suppliers can align their cultures and develop strategies to take advantage of efficiencies in the market. Strategic partnerships are not sole-source awards, but rather longer-term agreements with indefinite quantities attached to the deliverables. Strategic partnering is closely analogous to ID/IQ contracts being used by some U.S. public-sector agencies.

U.S. Parallel – Pentagon Renovation

The Pentagon renovation project is employing new techniques that could change the face of government construction procurement. The process is being pioneered by Walker Lee Evey, the Pentagon renovation's top manager. Mr. Evey is a former official of the National Aeronautics and Space Administration (NASA) and was responsible for space station procurement projects. The renovation is the first major rehabilitation of the 58-year-old building, and is expected to last through 2011. There is no bidding on the project. Design-builders are hired through a best-value selection. Oral presentations are being emphasized, and written materials are being kept to a minimum. At least 50 percent of the evaluation is based on the offeror's oral presentation. A great deal of emphasis is being placed on the offeror's past performance. Hensel-Phelps won a best-value, fixed-price, design-build contract for a new building in conjunction with the renovation. The contract carries a "zero target profit." The offerors priced the project at no profit, but are incentivized through a combination of award fees and incentive fees. The award fee sets aside an established pool of money, typically 8 to 10 percent of the contract cost. The fee is paid at various intervals during the contract if the design-build meets certain criteria. The incentive fee is based on a 50:50 split between the contractor and the owner for cost savings and overruns.

PERCENTAGE OF DESIGN IN THE SOLICITATION

The amount of design contained in a design-build solicitation can vary greatly. As shown in the figure below, design provided by the owner can vary from –10 to 100 percent. A design may be considered –10 percent if it is driven by a developer and brought to the owner as an unsolicited proposal or it may be 100 percent complete if the risk for errors and omissions are "novated" to the design-builder through the contract terms. Most design-build highway construction projects in Europe had design content in the solicitation documents that ranged from 10 to 50 percent.

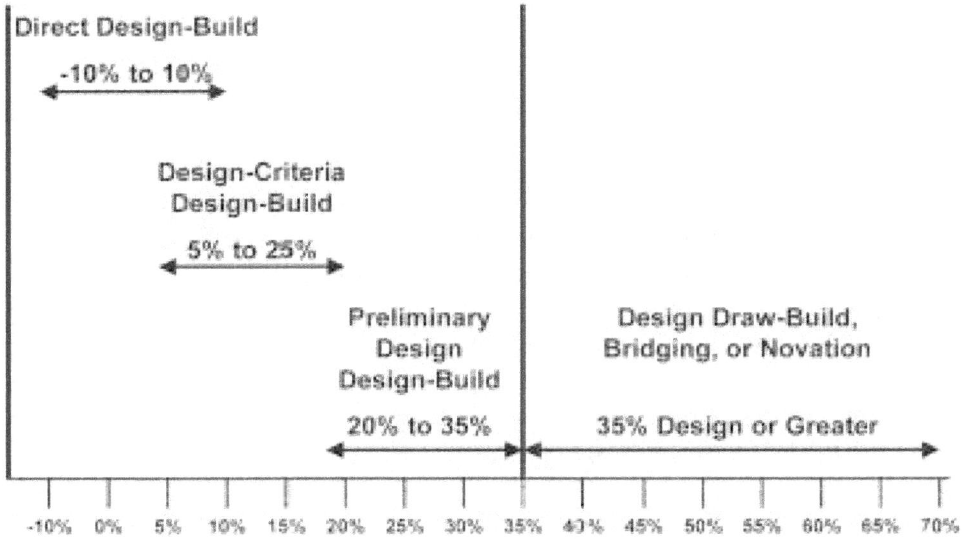

Percent of design in design-build solicitation.

The level of design contained in design-build solicitations was found to vary with the complexity and duration of the project. Relatively simple projects that could be completed in short periods of time contained more design in the solicitation, whereas complex projects or those with long-term operation and maintenance components contained less design. Those projects containing less design utilized outcome, or performance, specifications in lieu of higher levels of design.

In the United Kingdom, simple design-build projects contain relatively high levels of design (50 to 80 percent). The high level of design results from the requirements and length of the project-approval process. To obtain approvals, design must be taken to a very high level. All of the projects are developed to a point where the right-of-way and environmental clearance for the overall project are obtained before contract award. This fact has inhibited the use of design-build on small projects.

However, the Highways Agency has discovered that it may be overdesigning the plans if it wishes to truly benefit from the design-build process. It is attempting to involve design-builders earlier in the process and provide less in-house design prior to tendering. One such contract has been awarded to date under a strategic partnering arrangement. Payment during the initial phase will be made on a time and materials basis subject to budget limitations. The initial proposal included a lump sum price for final design and construction that will be adjusted for changes in design during the initial phase.

After analyzing the first eight DBFO projects, the British Highways Agency has discovered that the benefits of design-build will be better realized through earlier contractor involvement. The following finding is stated in its report, *DBFO – Value in Roads: A Case Study of the First Eight DBFO Road Contracts and Their Development*:

> The full potential of efficiencies, innovation and whole-life cost analysis inherent in the Private Finance Initiative is likely to be fully unlocked only when the private sector is involved in the outline design of the road scheme, which they are then obliged to construct, operate and maintain under a DBFO contract. This requires the private sector to assume some planning risk. Some of the DBFO projects announced introduce the concept of planning risk and will test the proposition that this will deliver better value for money.

The Dutch are much more conservative with the level of design in their design-build solicitations. In the past 10 years, the choice of design-build was in some cases motivated by the unusual nature of the project, in which as much as possible innovative power of the private sector was desired. Often, however, RWS had doubts about the design, resulting in disputes in which more and more responsibility for the design came back on RWS's shoulders. In extreme situations, this led to two separate tasks, as if a normal design and a normal construction contract were stuck together. This might still have the advantage that there is synergy between the designer and the producer, but it does not meet expectations connected with the design-build concept.

France and Portugal conduct design-build through the use of long-term concessions. Design-build is inherent in the concessionaire structure. Since the concessionaire is responsible for the long-term operation and maintenance of the facility, the government works almost exclusively on an outcome or performance basis. The design content in the solicitation is minimal because the whole lifecycle risk is being given to the concessionaire. With the risk, the owner transfers the responsibility of design to the concessionaire.

DESIGN AND CONSTRUCTION ADMINISTRATION

In Europe, as in the United States, some highways agencies are turning to design-build because of staffing shortages. Combining design and construction in one contract has the allure of decreasing the owner's staffing needs. Although design-build does allow for less owner staff, the profile of the staff also must change. The highways agencies no longer take the role of designer and inspector, but take the new role of definer of performance criteria and auditor of quality. As in the United States, European design and construction administration on design-build projects varies greatly from the traditional design-bid-build process.

As discussed in the previous section, the design-build is awarded with a varying level of design and differing levels of performance specifications. The countries that have been most successful with design administration after award of the design-build contract have taken the role of "reviewing" design rather than "approving" design. Because of the rigid approval process that RWS applied to the design-build process, disputes occurred over design development, leading to more and more responsibility for the design coming back to RWS. Design administration on design-build projects runs more smoothly when the highways agency allows the designer to take control of the details of the design. It is the owner's role to ensure compliance to the design-build solicitation and protect public safety. However, the owner must be willing to trust the design-builder with the details of the final product.

The United Kingdom also takes an approval role, but its system seems to have moved along more smoothly than that of the Netherlands. In the United Kingdom, the owner approves the design as the project progresses. The owner commits at 80 percent of the design development, but before construction documents are begun. There is a certification process, which will ensure that what is committed to is actually built. The government ensures that standards are satisfied. Owner change directives are directed and paid for by the owner. There is a procedure for creating, processing, and executing changes. Either the owner or the contractor can promote change. There is difficulty in pricing because the contract has already been awarded. Both parties must act in good faith.

The design-build process requires fewer differences in construction administration than it does in design administration. Once the project is under construction, a design-build project functions much like a design-bid-build project, with two significant differences: (1) payment methods and (2) QA/QC.

Design-build contracts typically do not allow for traditional unit pricing techniques because the design is often not defined enough to generate accurate quantities or even final unit price line item descriptions. The Europeans are utilizing both milestone payments and payments by percentage complete on design-build projects. These are not standard systems in U.S. highway construction and they will take time to develop. However, there are examples of these methods being used currently in the U.S. highway sector and they are standard practice in the U.S. building sector.

Design-build contracts typically utilize more contractor-controlled quality control processes because the design-builder owns the details of design. With contractor-controlled quality control, the highway agency takes on the role of quality assurance and quality auditor. This does not mean, however, that the owner totally gives up quality control. In the United Kingdom, the Highways Agency has the right to stop work if the contractor's quality plan is not being met or if the owner sees construction work is not proceeding as the owner requires. In the Netherlands, the contractor makes a quality plan, right after signing the contract. The contractor must be able to prove that it can ensure quality on three levels: product, production process, and quality system. The contractor is required to show certain documents from its quality assurance plan to the road owner. The road owner has the right to do checks/audits on all three levels. In Portugal, on the other hand, the owner does not have the right to stop work, but rather relies on long-term maintenance agreements with the design-builder/concessionaire to ensure quality.

A general trend was witnessed in Europe for the issue of quality in design-build contracts. Quality is being ensured through tying the design-build contract to finance, maintenance, and/or operation. Giving the design-builder responsibility for the entire lifecycle ensures quality in the constructed product. The use of alternative financing, operation, and maintenance, in conjunction with design-build contracts, minimizes the need for owners to perform time-consuming and redundant quality assurance roles. The difficulty lies in writing and enforcing appropriate performance criteria. The lessons learned relating to performance contracting are detailed in Chapter 5: Performance Contracting.

RISK ALLOCATION

Design-build contracts allow the owner to give more control to the contractor for third-party risks associated with utilities, environmental permitting, and right-of-way acquisition. The allocation of these risks varied greatly from country to country and even from project to project within each country. Each project has unique third-party risks and should be dealt with in an individual manner.

The United Kingdom is trying to transfer more third-party risks, where appropriate, through its PFI. New framework contract documents spell out the risk allocation between the owner and contractor. The scope of the "classic owner risks" that is allocated to the contractor depends on the point in the lifecycle of the project in which they negotiate the contract. Utility risks can remain with the owner or be transferred to the design-builder in appropriate situations. The owner retains the risk for land acquisition, as the owner must give the land as a precommencement condition to the contractor. Environmental risks are very constrained. The owner retains the risk for the environmental impact study, and all environmental risks associated with the construction processes are borne by the contractor.

The Netherlands is not as aggressive in transferring third-party risks as is the United Kingdom. The road owner generally takes third-party risks. If the utilities are undersized or mislocated, it is the responsibility of the design-builder. Responsibility for the time delays resulting from undersized or mislocated utilities, however, is uncertain. The general policy is that the party best able to manage the risk should handle it.

In Portugal and France, the concessionaires have a very large stake in third-party risks. Since the contract terms are so lengthy, and financial ownership of the asset is being transferred to the concessionaire, the majority of the third-party risks also are transferred. Portugal has even had to transfer some of the environmental permitting process to the concessionaire to meet its aggressive construction plans. However, Portugal would choose to retain this risk for financial reasons whenever possible.

USE OF WARRANTIES

Several U.S. agencies have sought to ensure design-build quality through long-term warranties. The use of longer-term warranties on traditional projects was prevalent in Europe in the early 1990s, as was found throughout the CATQUEST scan tour. Since that time, long-term warranties have been replaced with long-term maintenance agreements. The scan team was not given details of the warranty terms between the design-builders or concessionaires and their subcontractors from the European highways agencies, but 5- to 10-year warranties are common on certain items. Again, the use of warranties is not required with long-term maintenance contracts because the contract itself is essentially a means of warranting the work.

In the Netherlands, RWS detailed two kinds of warranties: (1) in-contract warranty for bankruptcy and (2) post-contract for nonsatisfaction. In traditional contracts the latter kind of warranty—usually for 3 years—was not very effective. RWS is therefore hesitant about the use of warranty in design-build contracts. In its view, one essential item is missing in the warranty question: quality assurance. RWS believes that the quality assurance must be specified correctly through performance outcomes, and then there is no need for long-term warranties. For further discussion of warranties, see Chapter 5: Performance Contracting.

DESIGN-BUILD-OPERATE-MAINTAIN

The design-build contract in Europe is evolving from a transfer of responsibility for design and construction to a transfer of the whole lifecycle through the addition of operation and maintenance to design-build contracts. The British Highways Agency probably has the most experience with design-build of all the countries that this scan team visited. Design-build is the delivery method of choice for the Highways Agency. Its program has grown from one of simple design-build to DBFO. In its report, *DBFO – Value in Roads: A Case Study of the First Eight DBFO Road Contracts and Their Development*, a number of objectives are listed for utilizing DBFO:

- To ensure that the project is designed, maintained, and operated safely and satisfactorily so as to minimize any adverse impact on the environment and maximize benefit to road users.
- To transfer the appropriate level of risk to the private sector.
- To promote innovation, not only in technical and operational matters, but also in financial and commercial arrangements.
- To foster the development of a private sector road-operating industry in the United Kingdom.
- To minimize the financial contribution required from the public sector.

The primary lessons learned on the first eight DBFO projects completed in the United Kingdom are listed in the report as follows:

- DBFO contracts have accelerated the introduction of cost efficiencies, innovative techniques, and whole-life cost analysis into the design and construction of road schemes and the operation of roads (although the Agency had started to review these possibilities in the context of traditional methods of procurement).
- The full potential of efficiencies, innovation and whole-life cost analysis inherent in the Private Finance Initiative is likely to be fully unlocked only when the private sector is involved in the outline design of the road scheme, which they are then obliged to construct, operate and maintain under a DBFO contract. This requires the private sector to assume some planning risk. Some of the DBFO projects announced introduce the concept of planning risk and will test the proposition that this will deliver better value for money.
- The risk allocation on DBFO contracts has been encouraging. Two areas where transfer of risk to the private sector has delivered good value for money are protestor action and latent defect risk. The Agency will continue to look for risk transfer to ensure that the DBFO contract remains off-balance sheet.

- DBFO contracts have delivered value for money. Cost savings (compared with the public sector comparator) have ranged from marginal to substantial; for Tranche I and 1A DBFO contracts, the average cost saving is 15 percent.
- Use of a model contract as the basis of negotiation for each DBFO contract saves bidders time in preparing their bids and provides significant efficiencies for the Agency, both in negotiation and in operating the contracts. The updating of the model contract is welcome, as it will reflect changes to provisions arising from negotiation.
- Training in negotiation for project teams and dissemination of accumulated knowledge on DBFOs and the Private Finance Initiative, generally, within the Agency continues to improve the quality of DBFO projects delivered.
- When devising the payment structure, the contracting body should determine what its objectives are for the service being provided, and the payment mechanism should be designed to incentivize the private sector to achieve those objectives.
- With eight contracts let and expressions of interest received for further projects, it is clear that a road-operating industry is developing. The same consortia (with a few changes in composition) have appeared as bidders on projects within each group.

The addition of operation and maintenance to the design-build contract solves the problems of design administration, construction administration, quality control, and use of warranties. However, the drafting and enforcing of operation and maintenance performance criteria creates new issues that are not commonly dealt with in most highways agencies in both Europe and the United States. Performance contracting is discussed in Chapter 5: Performance Contracting. The combination of financing with DBOM contracts is discussed in Chapter 7: Concessions.

SUMMARY

The European countries visited on the scan tour have used design-build delivery on a much more extensive scale than has the United States and provided the scan team with many valuable insights. Direct design-build contracts are employed between highways agencies and design-builders. Additionally, design-build contracts are inherent in PPPs and concession contracts. The primary lessons learned on this scan tour relate to the types of projects utilizing design-build, the use of best-value selection, the percentage of design in solicitation, design and construction administration, third-party risks, the use of warranties, and the addition of maintenance and operation to design-build contracts. U.S. highway agencies can apply these findings to many of their existing design-build contracting methods. The scan team recommends that the following concepts be explored in the United States as a means to speed the delivery of our infrastructure and increase the quality of construction and maintenance:

- Capitalize on best-value selection processes to promote competition and innovation among design-builders.

- Promote appropriate use of performance specifications with low levels of design in design-build RFPs to promote innovation and accountability from the private-sector proposers.
- Assign third-party risks to the party in the contract that can best control them.
- Ensure construction quality and cultivate a pool of qualified lifecycle service providers through the incorporation of maintenance and operation into design-build projects.

Chapter 5: Performance Contracting

Performance contracting is in its infancy in the U.S. transportation sector, but the tools and techniques are well established in Europe. Performance contracting provides a builder or maintenance operator with performance specifications that must be met, by employing whatever means the contractor determines most economical. These performance specifications are then continuously measured against a set of performance indicators as a basis for payment. Performance contracts are thought to allow much more room for innovation through creative construction methods—lowering the overall price of a given project. Additionally, performance contracts necessitate alternative procurement and payment practices, typically utilizing past performance and end-product qualities as measurements.

The scan team discovered applications of performance contracts in Europe for term maintenance, design-build, DBFO, and concession contracts. Since design-build and concession contracts have previously been discussed in this report, this chapter will focus on term maintenance contracts. The Netherlands and the United Kingdom are utilizing performance contracts for term maintenance contracts, but France and Portugal employ concessions for long-term maintenance agreements. This chapter focuses on examples from the Netherlands and the United Kingdom.

In 2000, the Netherlands decided to include more innovative types of contracts as part of its market approach to procurement. Within a few years, about one-third of all contracts for construction and maintenance will be performance oriented. In most cases these contracts will be integrated contracts containing design, construction, maintenance planning, and maintenance execution. In special cases contractors will be selected on the basis of their design proposals. A more detailed discussion of performance contracting in the Netherlands is included later in the chapter.

The United Kingdom uses several forms of performance contracting. Since performance contracting on its DBFO projects was discussed in the previous chapters, the discussion in this chapter will focus on MACs used for term maintenance of the Dutch motorway and trunk road system. The United Kingdom started with 3-year maintenance contracts for a limited scope of work. Currently, the term is 5+1+1 (5 years as a base plus two 1-year options) if the provider, the contractor, is achieving the performance indicators successfully. The scope of work also has expanded from the initial concept. Emphasis is being place on integrated supply chain management. The selection process includes evaluation of the plan to provide goods/services and also risk allocation within the contractor team. Maintenance includes routine matters and limited reconstruction work. If reconstruction costs are above a specified level, the job is separately procured.

The essential lessons learned concerning performance contracting on this scan can be summarized into the categories of performance specifications, performance indicators, warranties, and QA/QC. The Dutch have developed a method of performance specification using five levels of specifications, which range from road-user wishes to requirements for basic materials and processing. The British have created a definitive set of performance indicators for measuring the performance of maintenance contractors in their MAC contracts. They also have also created a Performance Review

Improvement and Delivery (PRIDe) group to audit and ensure the integrity of the system. The length of warranties, along with the projects/products being warranted, is examined across all of the host countries later in this chapter. The chapter concludes with a discussion of how QA/QC is being employed differently in performance contracts.

PERFORMANCE SPECIFICATIONS

Performance specifications are perhaps the most critical elements of performance contracting. U.S. highway agencies have a long history of creating and maintaining extensive prescriptive specifications that detail the materials and processes for construction. Performance specifying for contracts involving maintenance and operations is not commonplace in U.S. industry. Performance specifying is a complete change in direction from prescriptive specifying because the main objective of the highway agency is to specify the performance level or outcome of the project and not the means and methods as to how that outcome is achieved. To transfer the responsibility for a design-build or maintenance agreement, the highway agency must allow the providers to create their own means and methods. This transfer of risk and responsibility is advantageous for the owner. If the design-builder or maintenance contractor provides the means and methods, they also are liable for the outcome. Again, there is a shift for the agency from provider of services to network operator.

In the Netherlands, the team observed a systematic approach to the drafting of performance specifications. The Dutch utilize a unique method of defining performance specifications in five levels of requirements, which range from road-user wishes to requirements for basic materials and processing. Performance specifications detail both the operating level and minimum condition of the facility at the time it is returned to public ownership. The following description of the process is adapted from a document given to the scan team by Mr. Arie L. Korteweg, MSc, Project Manager Quality Assurance, Ministry of Transport, Public Works and Water Management. It succinctly describes the Dutch method of performance specification in all types of contracts.

> The Netherlands National Public Works Department has recently decided to innovate its market approach considerably. Within a few years about one third of all contracts for construction and maintenance will be performance oriented. A list of 60 pilot projects has been scheduled. In most cases these contracts will be integrated contracts containing design, construction, maintenance planning and maintenance. In special cases contractors will be selected on the basis of their design proposals. The central division supports this change process by developing standard functional requirements.
>
> In fact the new approach will award more tasks to the private companies. They will have to bear more responsibilities and liabilities. In these situations the national agency branch offices will make careful decisions regarding the tendering procedure, the form of contract and the contract management. This should be done on the basis of risk analysis. More innovative behavior in this decision making process is being supported by our division. The contractor should have no longer obligations based on detailed technical prescriptions; instead only functional

contract requirements, describing the desired performance of the object, should be used.

Requirement levels

RWS tries to distinguish certain performance levels in specifications. In road construction and maintenance, five levels may be used:

1. Road-User Wishes – the car driver wants a usable road from A to B, which is safe and convenient.
2. Performance Requirements – road-user wishes are translated into requirements for the pavement surface, such as skid resistance, smoothness, noise reduction, and evacuation of precipitation.
3. Construction Behavior – to achieve performance requirements the behavior of the construction may be specified (e.g., elastic and plastic deformation, durability). Construction and materials may be freely chosen by the contractor.
4. Materials Behavior – such as elasticity, plasticity, fatigue, and compactability. The contractor can optimize materials in the specified layers.
5. Requirements for Basic Materials and Processing – current standard requirements are on this level.

For a good preparation of new contract forms, it is essential to distinguish these levels clearly and consistently, not only for the pavement but also for realization and maintenance of all other objects, such as green surfaces and hydraulic engineering structures. (This department is also responsible for the main rivers and canals.)

Relation between form of contract and requirement level (see table below)

There is a relation between the form of contracts and the levels of specifications, although not a straightforward (one-to-one) relation. Level 2, for instance, is usable in maintenance performance and DC(M) contracts. But when desired lifetime is longer than the contract time, there might be risks that make it unavoidable to go down to level 3. The contractor has to ensure future construction behavior.

Traditional Contracts:

Description of the work by activities or construction parts, with cost homogeneous quantities.

Quality specification on instruction level.

Maintenance Performance Contracts:

Description of the project referring to drawings; not quantities.

Quality specification on level 2 and, if lifetime is an important issue, level 3 or even lower.

Design-Build

Terms of reference on level 2 and, for lifetimes exceeding contract period, level 3.

Public-Private Partnership

Users level (1) and if necessary level 2 (for instance, because of total road network uniformity).

LEVELS OF REQUIREMENTS IN DIFFERENT CONTRACT FORMS.

Contract Form	Level 1 Road Users Wishes	Level 2 Performance Requirement	Level 3 Construction Behavior	Level 4 Materials Behavior	Level 5 Raw Materials and Processing
Traditional	----------→	----------→	----------→	X	x
Maintenance Performance	----------→	X	x	x	x
Design-Build	----------→	X	X	Contractor	Contractor
Design-Build-Maintain	----------→	X	Contractor	Contractor	Contractor
Public-Private Partnership	X	x	Contractor	Contractor	Contractor

Key to symbols in the table above.

Symbol	Meaning
X	In these contracts this will be the first level to think of.
x	In many cases these levels will be used for considerable parts of the project. A contract for a normal civil engineering project will always have a hybrid character: there will be always parts of the object that have to be specified on different levels than the main part for many reasons (X and x refer to the main part).
----------→	The arrows indicate that during initial preparation of a project one should always start with level 1, reasoning down to the desired contract level.
Contractor	The contractor will have to translate the contract specifications down to the instructions for his personnel, on level 5 or even lower.

The table above provides a general framework for the level of specification required for the various contract methods. The following tables provide more specific details on the technical work categories and the characteristics for each of the specification levels. The technical work categories are divided into construction field and maintenance field definitions. These levels must still be disaggregated further for individual project contracts, but the following two tables offer insights for the technical work category operation and the characteristics of the specifications.

CONSTRUCTION FIELD – DEFINITION SPECIFICATION LEVELS, CHARACTER OF SPECIFICATIONS.

Technical work-category	Level 1 Road Users Wishes	Level 2 Performance Requirements	Level 3 Construction Behavior	Level 4 Materials Behavior	Level 5 Raw Materials and Processing
Geotechnics	safety and convenience resulting from settlements, stability, influence on adjoining structures	settlements, consolidation time, stability, influence on adjoining structures	drainage capacity for consolidation, stability, deformations	materials, drain distance	
Groundwork	conformity with political alignment decision	settlements, consolidation time, stability, influence on adjoining structures	dimensions according to design, permeability, stability	general properties, embankment materials	% fines, compaction
Road Drainage	-	drainage pavement surface	discharge	drains watertight, durable	
Pavement Base	-	-	bearing capacity, crack distance	durable	particle sizes, adhesive agent, compaction
Asphalt Pavement	section traffic speed, number and seriousness of accidents, noise complaints	smoothness, skid resistance, noise level, water on pavement, life time	smoothness, skid resistance, strength, fatigue, noise emission	smoothness, skid resistance, mix properties	particle sizes, bitumen quality, compaction
Concrete Pavement	section traffic speed, number and seriousness of accidents, noise complaints	smoothness, skid resistance, noise level, water on pavement, life time	smoothness, skid resistance, strength, fatigue, noise emission	smoothness, skid resistance	particle sizes, cement quality, compaction
Markings	road safety and drivers' satisfaction	reflection, degradation	reflection	durability	product and process certificate for paint, thermoplastics and their application
Crash Barrier	traffic safety	crash behavior, life time		durability	materials
Noise Barrier	political decisions	noise reduction, aesthetics		durability	according to drawing, materials
Green Surfaces	citizens' satisfaction	realization of landscape design	realization of specified greens quality		quality planting material, soil end planting methods
Shore Protection Mattresses	maximal ground surface loss, ecologically sound			permeability, sand-tighiness	
Stone Revetment	maximal ground surface loss, ecologically sound			stone sizes, mass weight, stone strength	

MAINTENANCE FIELD – DEFINITION SPECIFICATION LEVELS, CHARACTER OF SPECIFICATIONS.

Technical work-category	Level 1 Drivers and Politicians Wishes	Level 2 Performance Object	Level 3 Construction Behavior	Level 4 Materials Behavior	Level 5 Raw Materials and Processing
Asphalt Pavement	section traffic speed, number and seriousness of accidents	Smoothness, skid resistance, water on pavement, life time, traffic disruptions from maintenance	smoothness, skid resistance, degradation, cracking, traffic disruptions from maintenance	smoothness, skid resistance, mix properties	particle sizes., bitumen quality, compaction
Concrete Pavement	section traffic speed, number and seriousness of accidents	Smoothness, skid resistance, water on pavement, life time, traffic disruptions from maintenance	smoothness, skid resistance, joints behavior, cracking, traffic disruptions from maintenance		materials and processes
Markings	traffic safety	reflection, skid resistance, durability, traffic disruptions from maintenance	reflection, skid resistance, traffic disruptions from maintenance	durability	product and process certificate for paint, thermoplastics and their application
Crash Barrier	traffic safety	crash behavior, life time, traffic disruptions from maintenance		durability	materials
Mowing		according to green management plan			mowing height and removal of cut grass
Pruning of Trees and Bushes		according to green management plan			weather conditions, wound treatment, removal of waste
Rubbish Removal	traffic safety, environment pollution	traffic safety, environ ment pollution, convenience of parking lot users			
Concrete Repair		aesthetics, protection of reinforcement bars, durability	aesthetics, protection of reinforcement bars		according to process certificate
Stone Revetment	maximal ground surface loss, ecologically sound			stone sizes, mass weight, stone strength	
Dredging	unrestricted navigation	maintenance of certain depths		dredging quantities of solid materials	

Even a casual examination of the specification characteristics above reveals a significant change from traditional prescriptive specification methods. As the risk for operation and maintenance is shifted to the private sector, so is the responsibility for creating the final prescriptive specifications. The agency's goal is to write comprehensive outcome definitions, which allow for maximum innovation from the private sector without compromising long-term quality or safety. In fact, the responsibility for quality and safety also becomes the responsibility of the maintenance contractor. Quality and safety are achieved through a process of continuous improvement as benchmarked against performance indicators established in conjunction with the performance specifications.

PERFORMANCE INDICATORS

Performance specifications can only be used successfully if the outcomes are *measurable* and *verifiable*. Each performance specification must have a set of performance indicators associated with it so that the highway agency can measure and verify the quality and execution of the product. In essence, the highway agency creates the performance specification and then audits the performance via the performance indicators. Again, there is a shift for the agency from provider of services to network operator, in this case through the auditing system.

The United Kingdom provided the scan team with an extremely comprehensive set of performance indicators for its MAC contracts. The system begins with a general Performance Requirements describing the desired outcome of the product from the owner. It then defines Key Performance Indicators, which describe the targets used to measure the performance requirements. These are further defined in the Area Performance Indicators, which enable the specific targets to be set. Finally, the system provides a Mechanism for Setting the Targets. The figure below graphically depicts the process.

To further clarify the process, definitions for these terms are given below, followed by actual examples of each step in the following tables. The definitions and tables are provided verbatim from the British Highways Agency's Model Document MAC, *Performance Indicators, Annex 12*, Issue No. 3, Revision No. 1, British Highways Agency, London, England (British Highways Agency 2001). Only a few brief examples are provided for illustration in the tables. An entire set of performance indicators contains many more individual items, but it is much less voluminous than even the simplest set of U.S. construction specifications.

Key Performance Indicators (KPIs) – are published by the Highway Agency (HA) in connection with its strategic aims and objectives. The Managing Agent Contractor (MAC) assists in the collection and reporting of KPIs and HA collates the results and reports nationally. Overall relative performance in achievement of the KPI targets will be used as a measure of the MAC's performance to the extent that the MAC, through the Network Board, will assist and enable the Secretary of State to achieve such targets.

Area Performance Indicators (APIs) – are identified within the MAC Contract. The indicators have been developed to be common to the different forms of procurement for maintenance management. Services are to be carried out and the work is to be performed in such manner as will enable the specified targets to be met. Overall relative performance in achievement of those API targets identified as being primary responsibility of the MAC will be used as a measure of performance.

Target Setting – except where network wide targets are specified within the Contract, setting of targets for APIs is to be agreed via the Network Board. However in the event of failure to agree, a mechanism by which targets will be set is included within the MAC Conditions of Contract.

The table below describes the Performance Requirement for Emergency Response. This is only one example of a Performance Requirement. Numerous Performance Requirements are actually used to describe the entire project—much like an outline specification would be used at the conceptual design stage in the U.S.

EXAMPLE OF PERFORMANCE REQUIREMENT TABLE.

Objective	Performance Requirement	Response to Incidents	Measurement Basis
Emergencies are responded to quickly and effectively and assistance is given to emergency services as appropriate to minimize danger, delay and disruption	A suitably qualified member of staff shall be on standby 24 hrs a day 7 days a week. Respond to all incidents as quickly as practicable, but in any event within the maximum response times. Attend the scene of any emergency affecting any element of the network and assist the emergency services as necessary in order to minimize any danger, disruption or delay to the public and pollution of watercourses or groundwaters into which the road surface water drains. Attendees shall be suitably equipped and trained to assist with the incident and shall provide relevant information about local drainage systems, outfalls and soakaways as necessary. The response team clears the highway and carries out all necessary cleaning and proposes repairs of the network following an incident that requires attendance under emergency conditions. In the event of any spillage, appropriate action, as directed by the emergency services or by the Environment Agency, shall be taken to prevent the escape of pollutants. The Provider tests and classifies all waste material arising from an incident, and contains, stores and disposes of all inert, industrial and non-hazardous waste material arising from an incident. In the event of a breakdown of the communication system, the response team shall assist the communication contractor (to be defined) by: • providing traffic management as required • testing, repairing and checking any electrical systems that are the responsibility of the Provider	0700 – 1900 within 1 hour 1900 – 0700 within 1.5 hours	Emergency Response Audit

The following table provides examples of Key Performance Indicators for Customer Satisfaction and Road Traffic Accidents. Along with a description of the Key Performance Indicators, the table provides for a target type, target value and the data source that will be used to gather those targets. As this table is provided from model contract documents, the target values are left blank. These target values are established on a project-by-project basis as described in the two tables that follow. Again, these are only two short examples used to illustrate key Performance Indicators.

EXAMPLE OF KEY PERFORMANCE INDICATORS.

Key Performance Indicator	Description	Target Type	Target Value	Data Source
Customer Satisfaction	Number of complaints per month raised through Highways Agency Information Line (HAIL) and Road Users' Charter Unit (RUCU), measured under: • Poor traffic management- Peak time closures • Litter and appearance of the network • Poor carriageway condition • Poor user facilities • Poor advanced signing of roadworks and diversions	Area Target (max)	[ ] number	HAIL database of all public contacts reported on a monthly basis
Road Traffic Accidents	(a) Accident rate in PIA per 100 vehicle km	Area Target (max)	[ ] rate	STATS 19 forms accident data acquired and reported by the Provider
	(b) Severity ratio measured as ratio of severe accidents to total accidents	Area Target (max)	[ ] rate	

To further define the key performance indicators, Area Performance Indicators are employed. The Area Performance indicators refine the responsibilities, performance targets and data sources used to actually measure the performance. The table below illustrates the Area Performance Indicator for Response to Emergency Incidents.

EXAMPLE OF AREA PERFORMANCE INDICATOR.

Area Performance Indicator	Primary Respon-sibility	Description	Target Type	Target Value	Data Source
Response to Emergency Incidents	MAC	(a) Percentage of incidents for which a response is provided within target time against total number of incidents reported.	Area Target (max)	[ ] %	RMMS Database, incident log and police records
		(b) Average time of appropriate response to incidents from the time at which notification of the initial incident is logged by the *Provider* for each road/ route.	Area Target (max)	[ ] minutes	

Finally, the system is based upon continuous improvement principles—namely setting a baseline and continuously improving upon this baseline. The tables for Key Performance Indicators and Area Performance Indicators leave space for setting target values on each individual contract. Defining the mechanism to set these targets is a critical element of the process. The table below illustrates how targets are set for each project.

EXAMPLE MECHANISM FOR SETTING TARGET

Period	Target Type	Mechanism for Setting Target
First 12 months of the Contract Period	Area Target (max) (measured value of the indicator is required to be less than the target)	Area Target shall be no greater than the measured value of the indicator for the Area Network averaged over each of the first 12 months of the Contract Period. Notwithstanding the foregoing criterion, the Area Target shall be not greater than 110% of the mean measured value of the indicator for each month of the relevant periodl averaged over each of the areas where such data are available within the whole of the trunk road network.
	Area Target (min) (measured value of the indicator is required to be greater than the target)	Area Target shall be no less than the average measured value of the indicator for the Area Network for each of the first 12 months of the Contract Period. Notwithstanding the foregoing criterion, the Area Target shall be not less than 90% of the mean measured value of the indicator for each month of the relevant period averaged over each of the areas where such data are available within the whole off the trunk road network.
Second and subsequent years of the Contract Period	Area Target (max) (measured value of the indicator is required to be less than the target)	Where the measured value of the indicator over the relevant period averaged over each of the areas where such data are available within the whole of the trunk road network is less than the measured value for the previous year, and where the Employer determines that a benefit accrues either to the Employer or to the road user as a result of the reduction in the average measured value, the Area Target shall be reduced by the amount of such reduction. Where there is no reduction in the measured value of the indicator as referred to above, the Area Target shall remain unchanged from its previous value.
	Area Target (min) (measured value of the indicator is required to be greater than the target)	Where the measured value of the indicator over the relevant period, averaged over each of the areas where such data are available within the whole of the trunk road network is greater than the measured value for the previous year, and where the Employer determines that a benefit accrues either to the Employer or to the road user as a result of the increase in the average measured value, the Area Target shall be increased by the amount of such reduction. Where there is no increase in the measured value of the indicator as referred to above, the Area Target shall remain unchanged from its previous value.

The performance indicators as described above allow for benchmarking and continuous improvement of performance contracts. In this manner, the government has transferred the *risk and the responsibility* for design and construction. The contractual terms define how the performance indicators are performed, but who measures this performance is the next question.

The United Kingdom is taking the posture to allow the contractor to measure and record many of these performance indicators. It then audits the results of these measurements in a rigorous manner through the use of the PRIDe group. The definition of the PRIDe team and its role is given below as taken from the MAC contract.

The Role of PRIDe

PRIDe operates outside formal contractual arrangements to establish and agree with the Project Sponsor a baseline audit and monitoring program of MAC.

Audits and Quality Management Systems

It is a contractual requirement for the MAC to operate effective, rigorous and comprehensive "first party audits" of their own activities and to operate under a certified system that requires regular third party audits. The baseline audit program performed by PRIDe as a "second party audit" is not intended to provide day-to-day verification of a provider's management activities. It would be serious cause for concern if PRIDe were to identify management or system issues that had not already been identified by the MAC.

Additional Auditing by PRIDe

In certain circumstances additional monitoring by PRIDe may be determined necessary by the Project Sponsor or may be requested by the NIAC and the Conditions of Contract include provisions for such additional audits to be carried out at the expense of the relevant service provider in the event of Quality Management System failures.

Performance Analysis and Benchmarking

In addition to the audit and monitoring role, PRIDe will analyze performance and establish benchmarks for future performance. This process will drive implementation of best practice and so deliver continuous improvement.

This method of setting performance indicators and auditing performance equitably transfers the risk and responsibility to the design-builder or maintenance contractor. The owner defines a desired outcome, requires the providers to measure their performance, and then audits the outcomes. The contractor has the ability to design the system that meets the cost, schedule, and scope requirements of the owner. The contractor is therefore accountable for the product quality, because it owns the design and construction methods used to achieve the performance requirement. Finally, there is no need for long-term warranties because the performance contractor is responsible for the project for a much longer portion of its lifecycle.

U.S. Parallel: Contractor Controlled QA/QC

In the United States, there is increasing recognition on the part of State and federal highway agencies that, on nontraditional projects (for instance, design-build), the QA/

QC programs require significant procedural revisions. More responsibility is being placed on the contractor, and highway agencies are assuming more of an audit role. This is very similar to the approach that Europe already uses. Utah, Arizona, and Colorado have already begun using this approach from several good-sized design-build projects (I-15 in Utah, Legacy Highway in Utah, Superstition Freeway in Arizona, and the Southeast Corridor in Colorado) and Washington State is using this method on a smaller design-build project (SR500 Thurston Way Interchange).

WARRANTIES

CATQUEST, the previous contract administration scan tour, discovered the use of longer-term warranties in Europe than typically employed in the United States. The 2001 contract administration scan planned to collect data on the length of typical warranties and the types of projects being warranted. However, the team found that the use of DBOM, concessions, and performance contracting was more prevalent than the use of long-term warranties. Many of the concessionaire-subcontractor relationships contain warranties, and some of these have longer terms than typically found in the United States. Unfortunately, the scan team was not given information on these warranties because of the limited time with concessionaires and the proprietary nature of the companies.

The Netherlands was able to provide information on warranties for its traditional contracts. It is slightly different from the U.S. system. Normal warranty periods for the entire projects are 3 years. In cases of doubts during construction, this period may be extended. Beyond the warranty period, the designer and the contractor are only liable in case of faults that were not noticeable in spite of due surveillance. For traditional contracts the Dutch have found that warranties are not cost-effective.

The host countries visited on the 2001 contract administration scan were not focusing their alternative contracting techniques on warranties. From what was witnessed, the United States is developing more comprehensive long-term warranties than those being used in the European host countries. The host countries are focusing their efforts on tying maintenance performance contracts to construction in lieu of long-term warranties through performance methods similar to those presented in this chapter.

QUALITY CONTROL/QUALITY ASSURANCE

An item of concern in performance contracting in the United States is QA/QC. In the United States, traditional QA/QC roles and responsibilities are not effective with performance contracting. Performance contracts observed during the scan tour placed the responsibility for QC solely with the contractor, and the owner retained only a minimal QA audit role. However, there is use of "stop" or "control" points on projects as a means for owner verification testing at critical points.

The Netherlands also is employing a unique process for quality audits in lieu of heavy owner inspection. The Dutch do this through a system of penalty points. Akin to the referee in a soccer match, the owner gives the contractor yellow or red cards for quality violations. One yellow card is a warning; two yellow cards, or one red card, mean that the contractor must stop work until the violation is remedied. The following is a summary of the Dutch QA/QC system for the high-speed rail line earthwork and bridge contract.

Yellow and red cards keep HSL under the budget

In order to obtain the high-speed line at not much more expensive than planned, the project bureau works with yellow and red cards. A red card stops the payments to the contractor. Supervision is concentrated on process management by the contractor. In case of severe defects, where quality and/or security are affected, the project manager can show the contractor a yellow card. If the problem has not been solved within the agreed time span, a red card follows and all payments in the sub-project involved are stopped. As far as known in the Ministry no cards have yet been used.

In the Netherlands, the project bureau is first to use yellow and red cards (as in soccer) in the construction field. In this way it wants to keep the design-build contracts under control without a large supervision organization. This application of the so-called Brussels model is preferred over the Bahamas-model (no interference at all) and the construction -site-model (intensive supervision).

The project bureau has learned from experience of Rijkswaterstaat's Construction Division, the Project Organization Betuweroute (freight rail track Rotterdam-Germany) and the Project Organization Westerscheldt Tunnel. The construction of the high-speed line from Amsterdam to the border with Belgium is a huge project, which has been subdivided into several parts and then tendered. The bored tunnel under the Green Heart is under construction and the contracts for earth works and bridges have been signed. That comprises 1 billion and 4.4. billion guilders. In total contracts of 7.7 billion have been signed. In addition to the system with the yellow and red cards the project bureau works with fines and bonuses. By including strong financial incentives the contractors are stimulated to advance completion.

The process outlined above constitutes a significant change from the traditional U.S. owner-specified QA/QC programs. The owner is entrusting the contractor to ensure the quality of the end product, and the owner ensures this through an audit process. The owners are conducting much less frequent verification and testing. This would be a significant change for U.S. owners, but it is not unprecedented. The key to the system involves setting appropriate performance indicators and then monitoring them throughout design and construction. The audit system will not be effective if there is not a mechanism for the owner to "stop work or withhold payment" for an unacceptable audit. This result also can be achieved through an incentive/disincentive plan.

SUMMARY

The scan team discovered applications of performance contracts in Europe for term maintenance, DBM, and concession contracts. Performance contracting provides a contractor with performance specifications or requirements. These requirements must be met through means and methods determined by the contractor. These performance requirements are then continuously measured against a set of performance indicators as a basis for payment. Performance contracts are thought to allow much more room for innovation through creative construction methods, thus lowering the overall price of a given project. The essential lessons learned on performance specification on this scan can be summarized into the categories of *performance specifications, performance indicators, warranties,* and *QA/QC*. The scan team recommends that the following concepts be explored in the United States as a means to speed the delivery of our infrastructure and to increase the quality of construction and maintenance:

- Catalog those performance contracting methods currently in use in the U.S. transportation industry.
- Explore the formation of an audit group, similar to the U.K. PRIDe group, as a means to nationally benchmark performance indicators for use by all States. This team will be able to ensure, through diligent benchmarking, that projects are being delivered at competitive costs in lieu of ensuring competitive costs through the current low-bid system.
- Employ an aggressive pilot study program to explore the use of performance contracting for both construction and maintenance to determine the efficiency of our current methods and to develop consistent and objective performance indicators. This pilot study will allow for the measurement and verifiable benchmarking of the performance and a trial of other promising performance contracting methods.
- Based on the pilot studies and other sources, create consistent performance specifications that define the owner's performance objectives, which can be used to promote consistency in specifications while allowing for innovation in design, construction, and maintenance.
- In conjunction with the performance specification system, the performance indicators identified in the pilot studies should be used to create a system of continuous improvement for the industry.
- Nationally benchmark the performance of long-term warranties against the use of performance contracts to determine which system provides better value to the public.
- Promote the U.S. trend for contractor-controlled quality control programs and develop incentive/disincentive systems for quality such as the red card/yellow card system used in Europe.

Chapter 6: Alternative Financing

In the United States, we consider the *pay-as-you-go approach* to be our traditional project financing method. The pay-as-you-go approach uses a combination of federal grant assistance and State revenues to pay contractors. The traditional system has functioned adequately through the years, but the rapid expansion of infrastructure needs in relation to the available funding has created a desire for alternative financing methods. More specifically, alternative funding sources and alternative payment mechanisms have the potential to improve the speed of delivery and the quality of our national infrastructure.

European highways agencies are taking a "whole-life" approach to financing, designing, building, and maintaining their national infrastructure. They are asking the private sector to compete for the entire project, including financing, and incorporate payment methods as an integral part of the plan to ensure construction quality, road maintenance, road availability, and safety performance. A widespread use of concessions is found at one end of the spectrum, with a virtual transfer of ownership of the asset to the private sector for 15 to 30 or more years. In less aggressive strategies, PPPs are being formed to overcome sociopolitical barriers without transferring the burden of long-term financing or maintenance to the private sector.

This chapter discusses alternative financing through an examination of alternative funding sources and alternative payment mechanisms. A series of PPP case studies is presented, but the more aggressive method of concessions is left for discussion in later chapters. The scan team found the use of concessions and performance contracting to be so prevalent that this report dedicates entire chapters (Chapters 5 and 7) to those subjects. It should be noted that many European countries are aggressively employing concessions and performance contracting in most of their long-term strategic plans.

Two primary differences must be considered in the discussion of European alternative financing. First, the countries visited on the scan have a very limited amount of taxes dedicated to transportation needs. This situation means that gasoline taxes and other transportation-generated revenues are not earmarked for transportation projects, but are put into a general fund with other taxes. The general funds provide money for a variety of needs, including transportation projects, in amounts based more on the political priorities than on the source of the revenues. The second difference is that European governments do not have the ability to use tax-exempt financing for public transportation projects, as is the case in the United States Although this means that interest rates are higher for some European projects, it also means that such projects are not subject to the management contracting rules applicable to U.S. projects financed with tax-exempt debt. (Internal Revenue Service [IRS] rules require contracts for projects financed with tax-exempt bonds to be structured within safe harbors that may limit the operations/maintenance term or the form or amount of contractor compensation. See IRS Revenue Procedure 97-13, and Sections 141(b) and 145(a)(2)(B) of the Internal Revenue Code of 1986, as amended.) The absence of tax-exempt financing in Europe makes private financing much more competitive with public financing. For example, in the United Kingdom, the interest savings realized when using publicly guaranteed funds, in lieu of private funds, is sometimes less than 1 percent.

FUNDING SOURCES

Alternative funding sources in Europe include a combination of bond and bank financing. Private financing is much more commonly used than in the United States In some cases, governments need to use alternative financing because they have reached ceilings for public debt. In others, private financing is simply the best solution for the project. No matter the motivation for the use of alternative funding sources, private financing creates a contract administration environment that is substantially different from that when financing is obtained from the public sector. Particularly where bank financing is used, if repayment is dependent on project revenues (whether in the form of tolls or other user fees including shadow tolls), the lender will want to play an active role in contract administration. The entire contract administration lifecycle is affected, from programming and design to administration and quality control. If the United States wishes to employ alternative financing and continue to deliver a quality product, all of these elements must be considered.

The primary funding mechanism in Europe is similar to the traditional U.S. pay-as-you-go system. Most of the national investment in surface transportation infrastructure is funded through the annual budget allocation process, as in the United States For instance, in the United Kingdom money is budgeted on a 3-year cycle and is appropriated annually. The Treasury funds municipalities separately. The United Kingdom also was the only country visited that has a tax dedicated to highways. In 2000, an act dedicated a portion of transportation taxes to transportation projects, but this has not been a popular concept with the Treasury and it is unclear whether it will continue. The balance of the investment in the national highway system in the United Kingdom is funded with bank financing through the use of concessions or PPPs.

The European Investment Bank and Public-Private Partnerships

The European Investment Bank (EIB) was created by the Treaty of Rome in 1958 and is the EU's financing institution. The shareholders of the EIB are the 15 Member States of the EU. The core objective of the EIB is to support a balanced development of the EU; and the majority of loans are made to projects in regional development areas. The EIB has a special focus on PPPs and PFIs. A number of the projects in Portugal and England that were visited on the trip and described in this report were financed by the EIB.

The EIB operates by reviewing the project and credit quality of the proposals brought to it by public and private promoters. It then lends to the promoters, either directly or indirectly through banks. The majority of loans are funded on the capital markets. The indirect lending is principally for small projects.

Through the EIB, the EU has turned to PPPs as a preferred method of providing public infrastructure. The speed of implementation and the efficiency gains resulting from the cooperation are drivers behind this choice. One consideration of the EIB is the availability of private-sector capital, although this is not an important factor in all countries. What is more important is the benefit from private-sector efficiencies and the capacity to share and manage risks. The private sector, seeing the opportunities, has responded positively and is increasingly proactive in proposing PPP structures for individual projects or even entire regions.

> Guy Chetrit of the EIB summarized PPPs and their role in Europe (Chetrit 1999)[[**not in Ref List**]].
>
> PPP are an additional financial instrument to support capital investment in economic and social infrastructure for public services. PPP are valuable because of their positive contribution in management, cost effectiveness, quality of service—they complement scarce public funding. PPP offer a more flexible approach to ownership, risk sharing, organization and regulation. However, they require appropriate control by contract or regulation to protect the public interest.
>
> The ultimate objective of any project, including PPP, is to improve the overall rate of return and the public interest. EIB supports PPP as a generally efficient means of achieving this objective. We encourage flexibility in PPP frameworks and in structuring transactions so that the structure chosen is appropriate to the business conditions of each country and the particular sector concerned. Flexible structures and a medium to long-term view are, in our opinion, the best way to maximize the position impacts of PPP.
>
> Criteria for Public Private Partnerships include:
>
> - Economically viable for the public sector
> - Financially viable for the private sector
> - Appropriate risk and reward balance for public and private sectors
> - Public sector value for money
>
> Generally, projects, which are considered for PPP, should be socio-economically viable. The majority of public service projects and investments will, however, not be financially viable on a stand-alone basis. The public sector will therefore choose the appropriate form of participation to ensure that the project has the appropriate risk-reward balance to make it financially viable for the private sector. In order to make PPP a long-term feature, the private sector will require appropriate returns while the public sector will ensure that it is acquiring the service at an appropriate price.

The following table shows a variety of structures used in the EU, ranging from the 100 percent publicly funded road networks shown in the bottom right-hand corner to the 100 percent private user-financed projects. In between, there are a large number of alternative structures that have been selected in accordance with project characteristics. A number of the case studies described in this report were financed through the EIB.

INFRASTRUCTURE PROJECTS IN THE EU

Construction contract			Private sector-owned	Mixed Public / Private	Public Sector
	Fixed Price	User Fee	[illegible]		[illegible]
		User Fee plus govt subsidy	[illegible]	[illegible]	[illegible]
		No User Fee	[illegible]		
	Cost Plus	User Fee	[illegible]	[illegible]	[illegible]
		User Fee plus govt subsidy		[illegible]	[illegible]
		No User Fee			[illegible]

U.S. Parallel – State Infrastructure Banks and State/Regional Development Banks

Recent U.S. DOT legislation has allowed the States and territories to establish state infrastructure banks (SIBs). In 1995, the National Highway System Act initially authorized 10 States to establish such infrastructure banks to accelerate State-approved transportation projects by making loans and providing other forms of credit assistance. In 1997, the U.S. DOT Appropriations Act authorized all interested states to establish such banks. By 2001, approximately 40 States and territories had established SIBs; by August 2000, these banks had executed $749 million in loans and other credit instruments in order to advance $5.2 billion in critical transportation projects.

SIBs have been capitalized (i.e., initial deposits have been made) through a combination of federal and State funds. Initial deposits have been made from small amounts of federal startup or seed capital, as well as through the earmarking of a portion of regular federal-aid highway and transit funds. The process of earmarking is referred to as "capitalization" of federal grant assistance funds. In addition, States have matched federal deposits from their general fund balances and/or general revenues. In some cases, States have matched at the minimum required rate (generally 20 percent), while in other cases States have contributed significant amounts of funds beyond the minimum requirements. In addition, some States have decided to invest State funds in their SIB on an annual or continuing basis, thus increasing the resources available for credit assistance to transportation projects.

Given the variation in State contribution/investment policies, current SIB balances range from several million dollars to more than $1 billion. In addition, SIB operational policies—including lending guidelines—differ depending on States' needs, priorities, and political environment. Accordingly, the effectiveness of SIBs in leveraging resources and stimulating transportation investment will vary markedly from State to State.

In addition to the SIBs encouraged by federal legislation, a number of development banks have been established at the State and regional level. Some of these banks may be referred to as investment banks, others as economic development banks, and still others operate as government banks. Regardless of the terminology employed, the essential mission of these banks is the same: to ensure economic stability and promote economic development in their respective geographic areas, often with a related goal of growing the tax base and/or tax collections. A wide variety of economic incentives are offered: loans, credit assistance, tax incentives or abatements, donation of public property, and infrastructure improvements. Because of their broad developmental focus, these banks may not provide credit assistance targeted exclusively to transportation projects. In addition, since they are established to promote development in a given State or region, their geographic focus may be somewhat parochial/limited. In fact, one State development bank may develop a package of loans and incentives to lure a company from another, often neighboring, jurisdiction. Accordingly, development banks may not reflect either the regional coordination or perspective commonly seen in EIB operations, or the sharp transportation focus found in SIBs.

Sweden's Approach to Alternative Financing

A Swedish delegation joined the scan team in the Netherlands. Sweden has a much different approach to alternative finance and contract management. The Swedish government places (sells) its general debt—including debt used for transportation projects—in Japan. The benefit of this practice is that Sweden benefits from very low long-term interest rates currently being paid in Japan—less than 1 percent—and protects itself against currency risk with an appropriate hedging strategy. Sweden also allows local governments to accelerate approved transportation projects by arranging their own financing, and simply credits the localities' investment, without calculating interest, in the year that the project would have been completed without local government financing. This practice effectively allows local governments to make an interest-free loan to the central government if they wish to accelerate their projects. A more detailed review of the Swedish practice of operating their "DOT" as seven profit centers was beyond the scope of this scan, but it warrants assessment and evaluation in future studies.

Sijtwende Public-Private Partnership

The Netherlands provided the scan team with two primary examples of alternative financing. The first is a small project, Sijtwende, jointly developed between RWS, KWS Contractors, and Strucon Integrale Project Managers. Sijtwende is a combination roadway project/residential development.

In 1938, RWS planned to build a roadway connecting state routes A4 and N44. The state has owned the right –of way for the project for many years. For more than 20 years the project had met with public resistance, and construction could not proceed. The roadway is located in a historic and affluent portion of town. The citizens were concerned that the project, as originally proposed, would have had a detrimental impact on the environmental, noise, and aesthetic qualities of their community.

The project is finally being constructed pursuant to a PPP agreement among the government of the Netherlands and the local developer joint venture. A local developer conceived the idea of developing a residential community on the property owned by RWS, with the roadway constructed below grade in order to maximize the buildable land and the value of the homes. There are two tunnels, each 1,785 meters in length. The development includes 700 residences, including multistory buildings with apartments that will be sold to individual owners as well as townhomes. The tunnels run through the center of the project and are covered by a public park.

As of the date of our site visit in June 2001, much of the tunnel construction was already completed. The route is scheduled to open in 2002. The first phase of the residential development has already been completed, and all homes have been sold. As each home is sold, the proceeds allocable to land value are paid to RWS, and the proceeds allocable to the value of the improvements are paid to Sijtwende. The PPP agreement specifies how the purchase price will be allocated between land and improvements. The parties estimate that approximately 15 percent of the cost of construction of the tunnel will be paid by the land sales proceeds, which is approximately 35 percent of the additional cost to the government of building the project as a tunnel. The government was willing to enter into this arrangement notwithstanding the additional cost, since it enabled a long-delayed, critical project to proceed.

U.S. Parallel

A number of agencies in the United States have explored joint development opportunities for transportation projects. In the highway arena, there are a number of constraints on joint development. 23 U.S.C. § 111 requires property on highways that have received federal aid to be used exclusively for highway purposes (with certain exceptions). In addition, the management contracting rules promulgated by the IRS restrict long-term private operations of projects funded with tax-exempt financing. The underlying premise for these constraints appears to be a belief that public property should be reserved for public use, that public funds should not be used to benefit specific individuals or entities, and that public agencies should not exercise the power of eminent domain except for a public purpose. The FHWA does not currently have a formal policy on joint development, other than the rules based on Section 111. Its prior policy (PPM 90-5 in Chapter 7 of the *Federal-Aid Highway Program Manual*) has been rescinded and has not been replaced by any other policy statement or regulatory authority.

In contrast, the Federal Transportation Authority (FTA) is actively encouraging transit agencies to explore joint development opportunities for U.S. transit projects. Congress has specifically authorized use of grant funds to support joint development projects "which enhance the effectiveness of any mass transportation project and are physically or functionally related to such mass transportation project or which create new or enhanced coordination between public transportation and other forms of transportation, either of which enhance urban economic development or incorporate private investment including commercial and residential development" as well as "other non-vehicular capital improvements that the Secretary may decide would result in increased mass transportation usage in the corridor." See 49 U.S. Code 5309(a)(1)-(5) and (7). Under FTA's policy statement regarding joint development, the present value of the cash payment or revenue stream from the private development must equal or exceed the fair market value or the appraised value of the property used for the private development, taking highest and best transit use into account. See FTA publication, *Innovative Financing Techniques*, Chapter 3. It is unclear whether the Sijtwende project would have met those standards.

Joint development has been used for non-federal-aid projects by many State and local agencies, including for roadway improvements, parking facilities, pedestrian overpasses that improve access to shopping areas, and mall and beautification projects.

The *Project Development Guide* published by FHWA's Office of Real Estate Services includes the following examples of joint development projects (found in Chapter 14 of the guide):

- Freeway Park, built over I-5 in downtown Seattle, Washington, consists of more than 5 acres of plazas, waterfalls, pathways, lawns, play areas, and flower beds. Part of this park is directly over the freeway itself, and part is the top deck of a multilevel parking garage. In complex agreements between several participants, the State Department of Highways built the freeway and the deck that supports the park, the City of Seattle built the garage, private funds built the plaza and a high-rise office building, and the Seattle Park Department assumed responsibility for creating and maintaining the park itself.
- Many States have utilized smaller pieces of excess right of way or areas within interchanges in conjunction with local governments to develop small green spaces and playgrounds. Usually these are leased to the local government on a long-term basis, and involve funding from local sources for provision of equipment and maintenance.
- Multimodal facilities have been developed to mix the highway use and mass transit needs. Installation of rapid transit facilities within the right of way of a highway is one of the more common methods of doing this, as is the joint development of bus or rail stations in conjunction with highway right of way.
- Ride-sharing or carpooling facilities are often used with a mixture of funding sources to accomplish mutually beneficial goals. DOTs may acquire the land, and other funds may be used for development of the facility and its advertisement.
- Leased parking under highway facilities or on other available right of way space is under the "Airspace Leasing" provisions of 23 CFR.

Westerschelde Tunnel

The second example of alternative financing reviewed in the Netherlands is the Westerschelde Tunnel. The financing for the project is unique in that the government is providing the financing for the tunnel but, upon completion, the government plans to sell the tunnel to the private sector for operation and maintenance. The original plan was to have the tunnel constructed entirely through private finds. However, the project proved to be too risky for the private-sector financiers. The government decided to share the risk by paying for the construction and then selling shares in the project, as a going concern, to the private sector. Revenues will then be generated from the tunnel through the use of tolls.

The tunnel is a fixed link between Zeeuwsch-Vlaanderen and Central Zeeland. As early as the 1930s, there were initiatives to improve the crossing of the Westerschelde. Options for bridges and tunnels were considered, but given financial, environmental, and ecological issues, a tunnel was chosen. The tunnel is 6.6 km in length. At its deepest point, the tunnel is 60 meters below the Amsterdam Ordinance Datum. Two 11-meter tunnels will each have two lanes. The tunnel is being bored with two tunnel boring machines working 24 hours a day, six days per week for 27 months. The total project will take 30 months from design through construction. When complete, the tunnel will provide an ultimate capacity for 27,000 cars per day.

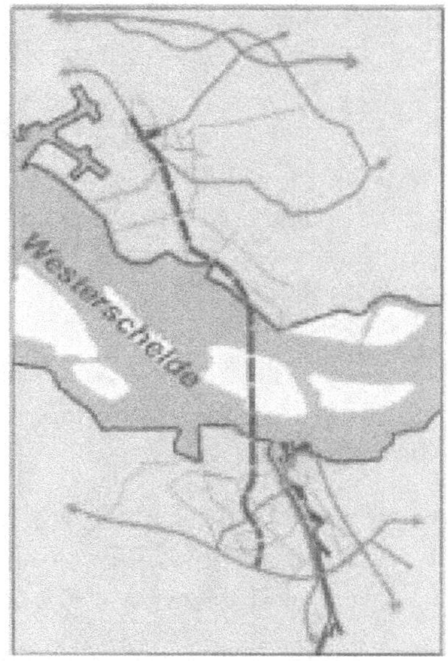

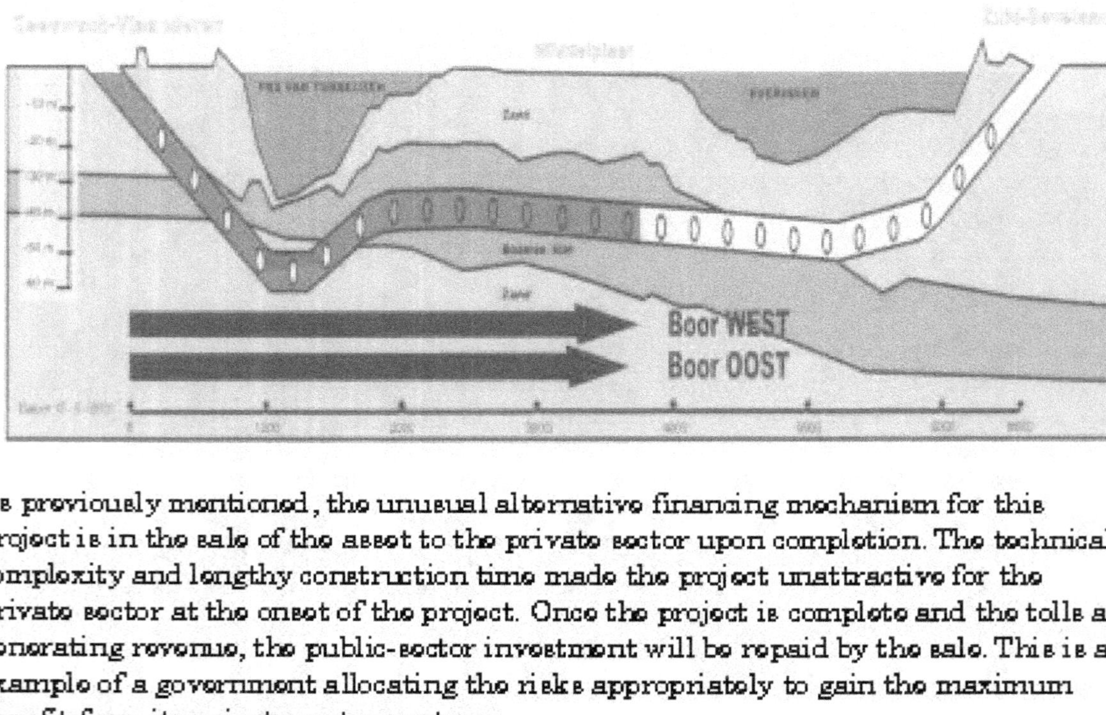

As previously mentioned, the unusual alternative financing mechanism for this project is in the sale of the asset to the private sector upon completion. The technical complexity and lengthy construction time made the project unattractive for the private sector at the onset of the project. Once the project is complete and the tolls are generating revenue, the public-sector investment will be repaid by the sale. This is an example of a government allocating the risks appropriately to gain the maximum benefit from its private-sector partners.

PAYMENT MECHANISMS

The use of alternative funding sources has led to a series of unique payment mechanisms in Europe. PPPs typically require the private-sector partners to carry the costs of design and construction much longer than traditional methods. Instead of being paid for the project on the basis of unit prices or progress or in a lump sum upon completion, the private-sector partners are typically considered temporary owners of the completed asset and receive payment over a much longer period of time. Common terms of payments in these ventures range from 15 to 30 years, but France awarded a 79-year concession for the A86 West Loop in 1999 (in effect through 2078).

Obviously, the contract administration implications of these long-term payment mechanisms are much different than traditional U.S. systems.

Many of the alternative financing payment mechanisms overlap with virtually every other section of this report, from contracting techniques to asset management. However, two particularly applicable payment mechanisms are discussed in this chapter. They both involve payment tied to usage, rather than tied to construction. The United States has experience in this type of payment mechanism through "real tolls" (or simply "tolls"). Tolls have been a successful alternative payment mechanism in various regions of the United States, but they have met with political barriers in other regions. Likewise, Europeans have faced political resistance to tolls. Two alternative payment mechanisms have come to the forefront: shadow tolls and active payment.

Shadow Tolls

Shadow tolls are an alternative financing payment mechanism in which the government pays a private-sector partner (PPP, DBFO, or concessionaire) for a project, based on the number of vehicles that use the facility. Traditional sampling methods and high-tech real count mechanisms are in use to count the vehicles for the shadow toll payments. The government receives the initial project financing from the private-sector partner, and the partner takes the risk/reward for the number of vehicles that use the road. In addition, the operational nature/characteristics of the shadow toll payments may assist the government in more effectively managing its debt because shadow toll payments are determined and made on a periodic basis—most commonly on an annual basis. Accordingly, the government and investment community may properly consider these shadow toll payments to be an item of operating expense; as an operating rather than capital expense, it generally need not be included in calculating debt ratios or debt capacity. Such an operating definition thereby provides the government with debt-management flexibility in the event that its revenues fall below expectations or if its cash-flow position deteriorates for some other reason.

Shadow tolls started in the United Kingdom under the DBFO Program. At the time of this scan, the British Highways Agency had brought eight DBFO projects to financial close and announced seven others (one in conjunction with the Scottish Office). The estimated capital value of the road plans within the DBFO program is just less than £1.3 billion (approximately US$2 billion). In cooperation with the Private Finance

Panel, the Highways Agency produced a case study of the first eight DBFO projects (DBFO – Value in Roads 1997). The following is an excerpt on the payment methods portion of that report.

> Payment is made for the provision of the road service. The Agency pays DBFO Company an amount, which is based on the number and type of vehicles using the road, with adjustments made for lane closure and safety performance. The payment mechanism was structured to meet Government policy objectives for the trunk road network and PFI requirements. It incorporates payment based on the following three criteria:
>
> > **Usage/demand** - shadow tolls involve payment per vehicle using a kilometer of the project road, in accordance with a tolling structure. They are referred to as 'shadow', as opposed to real, tolls because the Agency, rather than the road user makes the payment for usage. The shadow tolls increase over time in accordance with an indexation formula.
>
> Different payments are due for traffic within different traffic bands and dependent on the length of vehicle. Bidders were asked to bid the parameters of traffic levels for a maximum of four, and a minimum of two, bands, with the proviso that the top band—anything exceeding X vehicle kilometers per annum—must have toll levels set at zero to ensure that the maximum liability of the Agency under the DBFO contract is capped.
>
> Within each traffic band the bidders specified a toll for two categories of vehicle; long vehicles (over 5.2m—which, most importantly, includes HGVs) and short vehicles (less than 5.2m). There is no available method of differentiating between the weights of vehicles, and therefore length measurement was used as a proxy for weight.
>
> Bidders set the bands and tolls from their own assessment of traffic levels. Most bidders opted for four bands with the lowest band representing a cautious view of traffic and tolls within that band set at a level that would cover debt service requirements (but would not provide a return on equity). The adjacent figure shows a typical banding structure proposed by bidders.

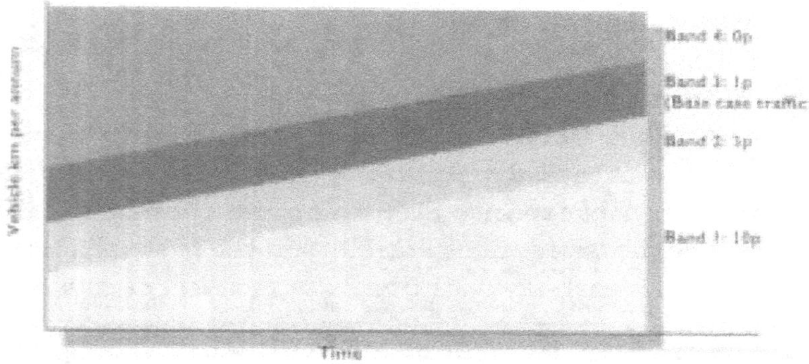

> > **Availability of service** - where the project road consists of an existing stretch of road with one or more construction schemes along its length, then (prior to the completion of any construction scheme) shadow toll payments will be made at a reduced level representing the cost of operation and maintenance

for the existing road. This level varies substantially depending on the nature of the DBFO project. In the case of the MI-Al project, which is virtually all new build, no payment is to be made until the Permit to Use is issued for the road to open to traffic.

Generally, once the Permit to Use is issued for a construction scheme and the road is open for traffic, DBFO Company receives 80% of the full level of traffic payment. When the construction works are completed and the Agency has issued the Completion Certificate, DBFO Company receives 100% of the traffic payment.

In most cases the toll payments step down again at the time when it is anticipated that the third party debt will have been fully repaid. This reflects the fact that revenue in excess of operating and maintenance costs at that stage is solely return on equity.

The adjacent figure shows a typical payment profile, assuming no variance in traffic or adjustment for lane closure or safety performance. The increase in payment over time during each 'step' results from indexation of tolls. Issue of the Permit to Use is marked 'A', issue of the Completion Certificate is marked 'B', and 'C' shows the point at which third party debt is anticipated to be repaid.

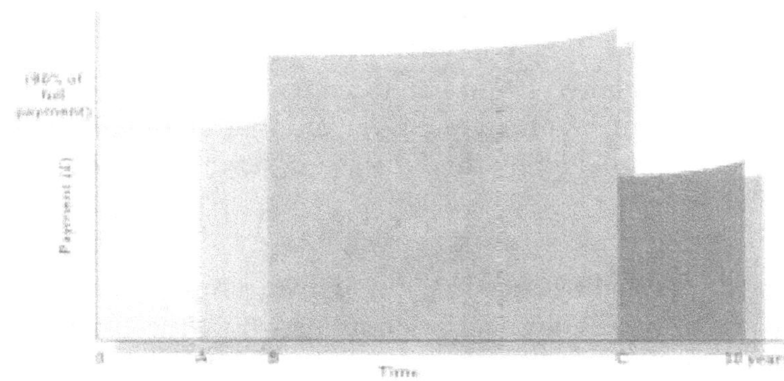

Performance - There are two aspects to performance payments: safety performance payments and lane closure charges.

One of the Agency's key objectives is to reduce accident levels on the trunk road network. In order to incentivize DBFO Company to address safety it is encouraged to suggest safety improvement schemes for Agency approval. If these improvements are agreed, DBFO Company constructs and pays for the scheme and is recompensed by receiving 25% of the economic cost of each personal injury accident avoided in the following five-year period. Accidents avoided are determined by comparing the actual statistics with data over the three years prior to the implementation of the scheme.

Disturbance and delay caused by lane closure is a significant performance issue to both the Agency and the road user. A deduction is made from the toll payment when lanes are closed. The size of the deduction is dependent on the number of lanes closed, the duration of the closure, the expected traffic at the time of closure (which encourages scheduled closure for maintenance at off-peak times) and the economic value of user delay, which can differ between business and leisure use. Lane closure charges are only made for closure within the control of DBFO

Company, for example no deduction is made for closures required by the police (in the case of accidents and emergencies) or utilities.

The Netherlands has used shadow tolls for two tunnel projects and has developed an interesting variation. The Highways Agency sees a potential problem in balancing the need to reimburse the private-sector initiative while meeting the agency's objectives. For the two shadow-tolled projects, the government made payments based on the number of cars that passed through the tunnel. It turned out to be very costly to the government. As another example, for projects where the amount payable for maintenance is fixed, the concessionaire would have a built-in incentive to keep traffic off the road. The Highways Agency now believes that payments should be structured to ensure that the agency's objectives are met. The agency's real problem is with congestion in the municipal community, which is not on the main system where shadow tolls are typically employed. For current projects the agency repays the private-sector investment using a modified shadow toll concept based on lane availability rather than lane usage.

For similar reasons, the United Kingdom also is experimenting with alternatives to shadow tolls. It has found that maintenance may not be optimized through the shadow toll system. The problem is that the shadow toll provider will be compensated for the number of vehicles using the system, even if the system is in poor condition. In essence, a shadow toll could still be paid if the vehicle passes over a poor section of road. Although there are incentives and penalties for maintenance quality written into the contract, the foundation of payment per number of vehicles still exists. One alternative that is being employed by the British Highways Agency is the AMPM.

Active Management Payment Mechanism (AMPM)

The AMPM is being developed by the British Highways Agency and has evolved from the shadow toll system. The shadow toll system has worked well for achieving many of the Highways Agency's goals, but it does not encourage active management of the project road to help reduce congestion and increase the reliability of trip times. AMPM consists of three main parts: congestion management, safety management, and service management. This section of the report describes the congestion-management element of the AMPM system. Yogesh Patel, who works with the Procurement Policy Division of the Highways Agency, provided the scan team with the following Summary of Principles for the AMPM system.

> Two of the Highways Agency's key objectives are to reduce congestion and to improve journey time reliability. The Congestion Management element is designed to do this by reducing payments for any times that congestion is experienced on the project road. It is considered that the DBFO Company can influence greatly the occurrence and levels of congestion through the effective management of the causes of congestion.
>
> #### DBFO Company Role in Managing Congestion
>
> The DBFO Company is considered to be in a good position to control and reduce congestion. The DBFO Company will therefore be required to accept the risk of predictable congestion such as roadwork, special events, slow moving vehicles, etc.,

and the risk of unpredictable congestion such as that due to accidents, poor weather, etc.

Management of these can be achieved through, for example: planning roadwork to be undertaken during off-peak times; liaising with the local authorities, police and other interested third parties to plan for impact of known events, having breakdown and response vehicles on standby; providing additional signing and break down vehicles during special events, placing temporary traffic management during emergencies, etc.

It is recognized that the DBFO Company will have limited control over things that cause recurrent congestion such as sheer volume of traffic demand above the nominal capacity of the road. The Payment Mechanism therefore makes allowances for this such that the DBFO Company does not carry this risk.

Basis of Congestion Management Payment

Tenderers will bid a fixed amount of money for each year of the commission. The amount bid will be divided up into weekly amounts and further divided for each carriageway section (approx 2km in length) and each hour of the day. The amount of weekly payment allocated to each section and each hour will be directly proportional to the expected level of traffic flow based on a rolling average of the number of vehicle kilometers traveled on each section of road for the corresponding hour of the day for the previous four weeks.

Full payment for each section and each hour will be made if'

- The road section satisfies minimum road condition criteria, and
- The target average speed for the road section is achieved.

Failure to meet the required minimum road section performance criteria will result in no payment being made for that road section for that hour. When the average speed falls below the target average speed, payments are reduced as detailed below.

Adjustment of Congestion Management Payment

The figure below demonstrates how payment will be affected by speed and flow with the key features being:

- At all levels of traffic full payment will be made if speeds are above 90 kph.
- Full payment will be made if traffic exceeds the deemed capacity of the road section, even if the speed falls below 90 kph.
- There will be graduation of the level of deduction for both speeds between 60 and 90 kph and between 80 and 100% of capacity.
- A bonus will be paid if flow exceeds 110% and speeds exceed 60 kph.
- The maximum bonus that can be earned is 20% of the payment for the hour and road section, if flow exceeds 120% of capacity and speed exceeds 90 kph.

Fundamental to the success of the new congestion management approach will be the need to give considerable freedom and flexibility to the DBFO Company to manage the flow of traffic. The DBFO Company will be expected to come forward with innovative proposals for managing congestion, some of which may need the Secretary of State to promote Orders. The Secretary of State will therefore give an undertaking not to unreasonably block DBFO Company's proposals. The grounds for objection by the Secretary of State will therefore be limited to the following:

- Has an adverse effect on safety;
- Has an adverse effect on the traveling public;
- Has an adverse effect on other transport networks/modes;
- Has an adverse effect on environment;
- Has an adverse effect on interested third parties;
- Is not consistent with the planning framework;
- Creates an unreasonable future liability for the Secretary of State; or,
- Is unlikely to achieve the intended outcomes.

The British Highways Agency is currently moving forward toward implementation of the AMPM system. This system incorporates the benefits of the shadow toll method and creates a new mechanism for providing better congestion management, safety management, and service management. The U.S. highway community should look at the advantages of this type of payment method and determine if it can be employed in U.S. markets to improve the overall quality of the highway system.

U.S. Parallel – Arizona DOT Active Payment Management Mechanism

Arizona DOT (ADOT) implemented a form of the AMPM concept on the State Route 68 design-build project. ADOT is using a contractual provision that requires the design-builder to measure speed consistency and performance through the 13-mile construction work zone. The contract provided a $400,000 travel time budget item that would be drawn against if the target travel time average is exceeded. Contractual incentives and disincentives would be implemented for performance above or below the contractual standard.

The design-builder elected to deploy an electronic license plate reader system developed by the British company - Computer Recognition Systems. This system uses a camera and a light source to capture license plate images of passing vehicles. The license plate number is taken from the picture by image recognition software, encrypted, and then sent to the central computer at the contractor's office through a high-speed data connection. There is a second camera at the end of the project, which takes a second picture, encrypts that license plate number and sends it to the central computer. The central computer then attempts to match up license plates that enter and exit the limits of the construction project.

The travel time incentive program is not very visible to the traveling public but they are still enjoying the benefits. Because of this contractual provision, the contractor has made great efforts to limit the delay to people traveling the corridor. The

contractor has made sure to limit the number of flagging stations throughout the project, and has scheduled work in such a way that it reduces the impacts to the public.

To date, the contractor has experienced very few times when the average travel time goal has not been met. The license plate reader system is able to match about 11 percent of license plates between the start and finish of the project. This rate is considered good compared with other license plate reading projects around the world and is adequate for the average travel time estimates used on this project.

For additional information, contact: Ron Williams, Arizona DOT (602) 712-7328, or William J. Higgins, Arizona DOT (602) 712-8274.

SUMMARY

European highways agencies are working closely with private-sector partners to finance and build projects much faster than traditional methods permit. Projects that are not viable using traditional funding mechanisms, either because of lack of funding or sociopolitical constraints, are being constructed through the use of PPP and other alternative funding sources. These alternative funding sources take a whole-life approach to project design, construction, and maintenance. As discussed in this chapter, the payment methods for these projects range from payment through proceeds of adjacent land sales owned by the government to real tolls paid by the users. As an alternative to real toll payment methods, alternative financing payments are being made through creative methods such as shadow tolls and AMPMs. These payment methods offer solutions that increase price competition from the private sector, but also incentivize and improve quality in the completed and maintained project.

U.S. highway agencies can directly apply these findings to many of their alternative financing methods. The scan team recommends that the following tools be explored in the United States as a means to speed the delivery of our infrastructure and increase the quality of construction and maintenance:

- Take a whole-life approach to planning through linking construction quality and maintenance to private financing.

- Leverage innovative concepts from the private sector to overcome social and political barriers for improving our infrastructure.
- Explore the use of shadow tolls and AMPMs as a means to defer infrastructure payments while decreasing congestion and increasing safety and availability.
- Measure and benchmark the performance of these alternative payment mechanisms to create an opportunity for continuous improvement measurements.

Chapter 7: Concessions

While only a minimal number of private transportation concessions have been awarded in the United States, some of the European countries visited on the scan are leveraging concessions for major portions of their highway systems. Concession agreements typically allow the concessionaire to design, build, and operate a project, with the right to receive revenues from operations and/or to receive payments from the public agency for use of the project. The use of concessions was found in each of the countries visited on this contract administration scan tour, and included both construction and maintenance of motorways. Concession periods vary, but were commonly found to be 3 to 5 years for maintenance and 15 to 30 years for construction and operation. In France, concessions have been an integral part of its program to develop, operate, and maintain its main highways for more than 30 years. Portugal is aggressively employing concessions as part of its strategic plan to develop its national highway system, and plans to have 90 percent of its national highway system administered by concessions by 2006. The United Kingdom has begun an aggressive DBFO plan as part of its national PFI, with commitments or plans for more than 15 projects to date. The Netherlands has embarked on limited use of concessions, primarily on tunnel and rail projects, and is experimenting with concessions for smaller maintenance and operations contracts.

The main reasons for using concession range from a lack of public funding to a belief that private financing and delivery provides a higher quality. Concessions also are being used as a means to provide benchmarks for public sector agency performance. As discussed throughout this report, many European highways agencies are beginning to take the role of network operator rather than provider of services, which leads to an outsourcing of production tasks through concession contracts.

Concession contracts can take many forms, and the definition of a concession contract can vary slightly from agency to agency. For purposes of this report, the definition of a concession contract will be taken from *A Draft Typology of Public-Private Partnerships* as written by Rémy Prud'homme for the French Ministry of Public Works, Transport and Housing (Perrot and Chatelus 2000):

> The concessionaire carries out all of the capital investment, operates the resulting service and is remunerated through service fees paid by users. The facilities are to be handed over to the oversight public authority at the end of the contract period.

Concession contracts are interrelated with alternative finance, as discussed in Chapter 6. Concessions are procured and administered using the contracting techniques described in Chapter 3 and have many of the same characteristics as the DBOM contracts described in Chapter 4. These contracts also implement the performance contracting techniques described in Chapter 5. Because the use of concession contracts is so prevalent in Europe, the scan team decided to devote a full chapter on concessions instead of trying to address them on a piecemeal basis.

The table below is provided to help clarify the differences in concessions and other PPPs. The options in column one of the table below provide the spectrum of PPPs from traditional agency management to complete privatization. The table was adapted from *A Draft Typology of Public-Private Partnerships* as written by Rémy Prud'homme (Perrot and Chatelus 2000)

DIFFERENT TYPES OF PUBLIC-PRIVATE PARTNERSHIPS

Option	Capital Investment	Operation & Maintenance	Commercial Risk	Asset Ownership	Contract Period	Discussed in Report
Public Agency Management	Public	Public	Public	Public	—	—
Service Contract (Performance Contracting)	Public	Public/Private	Public	Public	1 to 2 years	Chapter 5: Performance Contracting
Management Contract	Public	Private	Public	Public	3 to 5 years	Chapter 3: Contracting Techniques
Concession of Existing Network	Private	Private	Private	Public	5 to 30 years	Chapter 7: Concessions
Concession of New Facility (Build, Operate, Transfer)	Private	Private	Private	Public => Private	20 to 30 years	Chapter 7: Concessions
Privatization	Private	Private	Private	Private	Indefinite	Not Discussed

Since many of the contract administration features of concessions are discussed in other chapters of the report, this chapter presents successful concessions case studies and valuable tools discovered on this scanning tour. Specifically, the incorporation of concessions into strategic plans for road networks is discussed via profiles of the Portuguese and French approach to network concessions. The selection of concessionaires is discussed using examples of the Portuguese approach and the Public-Private Comparator utilized by the British and Dutch Agencies. The chapter concludes with a discussion of the duration of concessions and measures of performance to ensure quality during concessions.

CONCESSIONS AS A PART OF STRATEGIC PLANS FOR ROAD NETWORKS

Through establishment of a partnership between the public and private sectors, concessions can be an effective means of satisfying the strategic needs of highway transportation agencies. Transportation projects involve a very high level of capital investment combined with an extremely long period for recovery of this investment. Additionally, forecasting the rate of payback is extremely difficult because of the number of variables that affect use of roadways, including the variety of transportation options available to users (alternative routes, mass transportation, etc.). Given the critical need for transportation infrastructure to support movement of goods and people, and the lack of inherent incentives for the private sector to provide highways, one of the traditional roles of government has been the delivery of public highways. In some European countries, however, there is a belief that the private sector can provide the same services at a higher quality and lower cost. In other countries, the public sector is not capable of or is not willing to make the financial investment required to complete major infrastructure projects. These are just some of the reasons for use of concession contracts as a part of highways agencies' long-term strategic network plans.

Of the countries visited on the scan, France and Portugal are the most aggressive users of concessions. They view their concessionaires as an extension of the highways agencies. In France, only one of the nine concession companies is wholly privately owned. In Portugal, 90 percent of the strategic National Road Plan will be under concession by 2006. In these countries there is a belief that concessions will deliver a better value for each Euro spent. The following table lists the financial and political advantages to the government of using concessions. The table was adapted from *In Favor of a Pragmatic Approach Towards Public-Private Partnership* as written by Corinne Namblard (Perrot and Chatelus 2000).

Financial Advantages	Economic & Social Advantages	Political Advantages
• Easing of budgetary constraints • Optimal allocation and transfer of risk to the private sector • Realistic evaluation and control of costs	• Streamlined construction schedule and reliable project implementation • Modernization of the economy and improvement of services • Access to financial markets, combined with the development of local financial markets	• A new role for the public authority • Allocation and not "abdication" • Project stability

Given the advantages listed above and other realities of each country, Portugal and France have turned to concessions as the primary method for implementing their national road plans. The following sections outline the specific use of these concessions in each country.

The Portuguese Strategic Plan

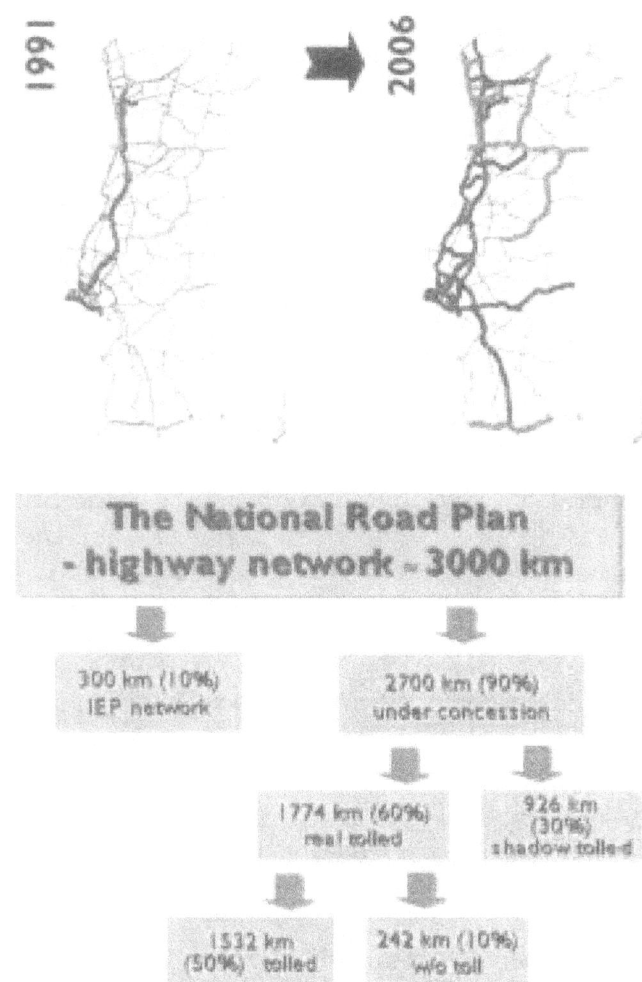

To meet its program goals quickly and efficiently, the Portuguese Highways Agency, Instituto das Estratas de Portugal (IEP), has made major changes in its methods of doing business over the past few years. In 1991 Portugal's roadway network included only 481 km of concessions. By 2006 it plans to have a total of 2,700 km of concessions in place—representing 90 percent of its national highway network. Use of a concession system is allowing Portugal to complete its strategic National Road Plan in 2006 – eight years earlier than the schedule using traditional methods.

The Portuguese use two primary payment vehicles for their concession contracts. The first means is real tolls such as those being used for U.S. concessions. Concessionaires finance and maintain the roadway in return for payments collected as tolls from roadway users. The second means of payment is through shadow tolls where the government pays fees to the concessionaire on the basis of the number of vehicles using the roadway. Shadow tolls are discussed in greater detail in Chapter 6: Alternative Financing.

The primary driver for the Portuguese concession plan was Portugal's entry into the EU and the need to strengthen its trading ability. The following paragraphs are adapted from Motorway Concessions in Portugal as written by Prof. Antonio Lamas, Director of the Portuguese Roads Authority (MES 1999).

> A bit of history is necessary to explain the present Portuguese "cycle" of motorway concessions. The first concession for the building of a tolled motorway network dates from 1972, with the creation of the private company BRISA to which it was granted. Shortly after the 1974 revolution, the majority part of the capital of BRISA was taken by the Government, and it became practically a public company. The rules originally set for granting a tolled motorway concession were simple but very exigent: it was required that an alternative route of good quality existed. This rule is still considered and is one of the reasons behind the decision of using shadow toll methods for the cases where no good alternatives exist and a motorway is the best immediate solution. Financially, the State budget could contribute to the initial investment up to 35% and to the economical equilibrium of the concession. The management of this concession was thus shared between the Roads Authority, representing the Minister of Public Works, and the Minister of Finances, introducing a style of regulation of the technical aspects separated from the financial matters, which still persists.
>
> Historically, Portugal has followed the other European countries in conceding emerging public services, with a peak during the last quarter of the nineteenth century. In the main cities, almost all utilities were concessioned to private companies. Some lasted almost a century: gas, electricity, telephones, water, trains and other public transports, etc. The nationalization of several concessions and the legal closing of most public services to private enterprise, was only terminated this decade, and the consequences of this policy in the Portuguese experience, or lack of experience, in the role of concessions was serious. One of the consequences was the generalized idea that the only source of funding for public services was the budget as a generalized effort of all taxpayers, and there was no adaptation to the modern thought of preferably calling the users to fund the required investments in new and modern utilities. It is a problem similar to that faced by countries that have a tradition of not tolling roads.
>
> For the new bridge over the Tagus River in Lisbon, Ponte Vasco da Gama, which opened for the Lisbon 1998 EXPO, a complex project finance concession scheme was devised. The same happened in the field of motorways: the limits of the concession of BRISA were redefined—that is, from being the concessionaire of "all" motorways, BRISA was limited to some main axis (with a length of 1180km) and privatized. BRISA still is one of the largest road concessionaires in Europe. At the same time, the present Government decided to open the sector to competition in order to complete, in a shorter period, the building of the planned motorway

network, which is essential for the equilibrium between regions and for the access to Europe (during the almost 20 years up to 1997, the sole concessionaire built only 680km). But it has to be said that the acceptance of the need for calling private funding into the financing of expensive roads was not accompanied by a corresponding development of the acceptance of tolled roads. And strong movements against new tolled roads have emerged all over the country: the Portuguese have accepted that the main axis that was part of the concession of BRISA could be tolled but not the new ones. Not because they are considered less important but because they were not expected and there has been insufficient discussion of the principles and usefulness of new concessions. This is not simply a Portuguese problem but it is necessary to take it into consideration in order to explain the context in which, for some of the new cases, the adoption of the DBFO system with shadow or virtual tolls (which is being translated in Portugal as SCUT) took place. It can be said that it was introduced in all situations in which a motorway was required and there was no good quality alternative, or the traffic forecast was not considered to be interesting enough as to bring sufficient competition between bidders.

In a concessions strategy such as that developed by the Portuguese, appropriate risk allocation is essential. The adjacent table describes the distribution for the responsibility associated with the risks of the strategic plan.

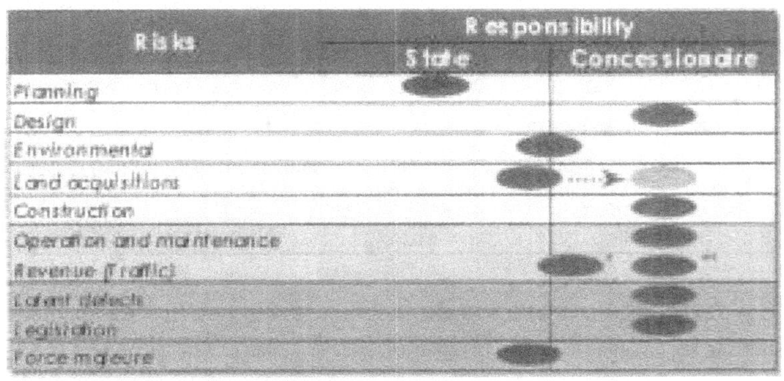

Risks
Responsibility
State
Concessionaire
Planning
Design
Environmental
Land acquisitions
Construction
Operation and maintenance
Revenue (Traffic)
Latent defects
Legislation
Force majeure

* SCUT Concessions (Shadow toll)
** Real Toll Concessions

The risk-control strategy suggests that the party best able to manage the risk bears it. For example, the risk of planning remains with the government. The risks associated with design, construction, operation and maintenance, latent defects, and legislation are assigned to the concessionaire, while there is a shared responsibility for environmental, land acquisition, and force majeure events. There is a shared risk for revenue in the shadow toll method. Planning is the only risk that the government maintains in full.

Two of the most difficult risks affecting transportation projects are environmental permitting and right-of-way acquisition. Based on current experience, the IEP's preference is to obtain environmental approval before launching its program, but it is not always politically possible. An alternative is to have the government retain the risk of failure to obtain approval. Many of the projects are subject to environmental problems that result in delay and thus a delayed commencement of tolling. When that happens the government compensates the concessionaire for additional design costs, additional consultant costs, and increased costs of environmental compliance, including land cost and cost of improvements. This situation can be extremely expensive.

Right-of-way acquisition cannot be totally delegated to the concessionaire because expropriation (condemnation) rights may be exercised only by the government. In practice, the concessionaires can participate in the acquisition process, doing everything up to the determination of need. The first Portuguese concessions gave the government primary responsibility for acquisitions, with parcels being identified by concessionaires and the government undertaking acquisitions. This method has proved burdensome. The most recent concessions have involved a transfer of significant right-of-way risk to concessionaires, by transferring more and more of the expropriation activities to the concessionaires. Concessionaires handle negotiations, and the government provides the public interest declaration. If contested, the matter goes to court and the government handles the case. The potential for delay in the court proceedings is a government risk. The government also bears the risk associated with any requirements to acquire property outside of the original corridor as a result of environmental approvals. In some cases, the government started acquisition proceedings early and ultimately discovered that the parcels were not needed for the ultimate project configuration.

The aggressive Portuguese concession program involves some adverse impacts. One concern is the loss of owner expertise under this program. In a period of less than 6 years, Portugal has moved from a program with 1 concession to a program with 14 concessions. This shift has enabled the IEP to downsize its engineering and administrative staff, but also has resulted in a loss of valuable expertise. Although the IEP has been relieved of direct responsibility for developing major projects, the IEP must continue to develop design, construction and operation standards, and policies that will be the basis for establishing the scope of the concessionaires' obligations. The loss of expertise will be felt for many years, both in the lack of resources for reviewing future concession proposals and in the administration and oversight of current contracts.

A Case Study: The Vasco da Gama Bridge in Lisbon

(adapted from Ferrot and Chatelus 2000)

The Vasco da Gama Bridge in Lisbon is one of the largest and most interesting concessions in all of Europe. Because of the geographical position amidst the inland sea created by the Tagus River estuary over its course, the length of the bridge was conceived to be approximately 12 km. When including all of the access roads and necessary motorway junctions, the total project comprised a length of over 18 km, with a price tag of some 6 billion francs (approximately US$883 million (including financing costs.

In spite of the subsidies provided by the European Union Solidarity Fund in the amount of about 2 billion francs (approximately US$294 million) (34.5 percent of total needs), the remaining financing required for the project exceeded the capacities of the Portuguese national budget's conventional financing system. For the complementary financing to be borne by the private sector instead of the national budget, the concession formula was chosen. To make this strategy viable, the contribution of subsidies proved essential. Looking at similar projects of the time (doubling the capacity of the Severn River Bridge and of the M25 tunnel under the Thames River in Dartford, England), Portuguese authorities came up with the idea of providing subsidies, of over 600 million francs in all, siphoned from revenues generated by the existing "April 25th Bridge", which had been run as a toll facility ever since its opening. The international tender for the new bridge's financing and construction as well as for the operations of both the new facility and the April 25th Bridge, all lumped into a single concession, was ultimately held in 1993.

In April 1994, the future concessionaire was selected; the contract was awarded to a consortium of construction companies (called "Lusoponte"), composed of:

- An English company, Trafalgar House, and a French company, Campenon Bernard SGE, each with a 24.8 percent share; and
- Five Portuguese companies splitting the remaining 50.4 percent.

The definitive concessionary contract was signed on March 24, 1995, with the primary stated objective being the new bridge's service startup prior to March 31, 1998, in time for the World's Fair.

In addition to the contribution of the Solidarity Fund, the plan of finance included an EIB loan for 2 billion francs and approximately 700 million francs in shareholder capital. The remaining amount was financed through a conventional bank loan.

The construction companies participating in this bold project needed strong international credentials in order to cope with the magnitude of the risks involved: financing and building one of the world's premier bridges within a span of 3 years, then assuming the risk of maintaining minimum traffic levels over the concessionary period, set at a maximum of 33 years, with the concession expiring once the number of crossings on both bridges combined has reached the 2.25 billion mark. The concessionaire was even given responsibility for the risk of right-of-way acquisition.

Despite all of the inherent uncertainties in a project of this scope, the Vasco da Gama Bridge was opened for service in accordance with specifications on March 29, 1998. Twenty-one months after its opening, the success of the Vasco da Gama Bridge has surpassed all expectations. Traffic in 1999 soared to 16 million vehicles, for a daily average of more than 43,000 vehicles. Even more extraordinary, however, is the fact that this success has not come at the expense of the older bridge, which reached a record level of crossings in 1999 with more than 53 million (for a daily average of 147,000 vehicles). In two years' time, the total number of river crossings by road has risen by 20 million, for an increase of nearly 40 percent, and this despite the opening of a rail link in 1999 between the two banks built on the April 25th Bridge's foundation. This explosion in the demand for cross-river transportation, subsequent to the significant increase in supply over a two-year span, provides an eloquent illustration of how a program can satisfy the pent-up demand of residents in a rapidly growing region.

The French Strategic Plan

The French have a long history of PPPs and concessions. In fact, the nation's first concession was granted to Adam de Craponne in 1554 for the construction of a canal. The Paris Metro is another concession. The City of Paris built the infrastructure, but it is run by a private company, Compagnie du Metropolitain di Paris (Group Empain).

Given this history, it is not surprising that the French utilize concessions for the majority of their motorways. The French use a real toll system with almost 100 percent of the cost for the roads being borne by the user. Tolls account for 65 percent of the capital motorway costs, with 19 percent from the government (3 percent for maintenance and 16 percent for building) and 16 percent from the local authorities (0 percent for maintenance and 16 percent for building). The tolls themselves are spent for financing (63 percent), toll collection costs (26 percent), and taxes (11 percent). The table below provides an estimate of the size and amount of motorways under concession.

Total Motorways (Jan 1, 1998)	9,309 km
Interurban Highways	8,319 km
• Toll roads	7,048 km
• Non-toll roads	1,271 km
Urban Highways	990 km

The French motorways system has steadily grown from 1,125 km in 1970 to more than 11,000 km in 2000. There are nine primary concessionaires, only one of which (Cofiroute) is fully private. Central and regional government bodies hold the remaining eight regional concessionaires through limited liability companies (SEMs). Some of the more profitable SEMs support the other less profitable ones. Some public companies have a private "firewalls" so they can compete with the private sector. SEMs are financed by the Caisse Nationale des Autoroutes (CNA), an autonomous public agency that raises the funds for highway construction. Private utility companies sometimes operate SEMs under short-term contracts.

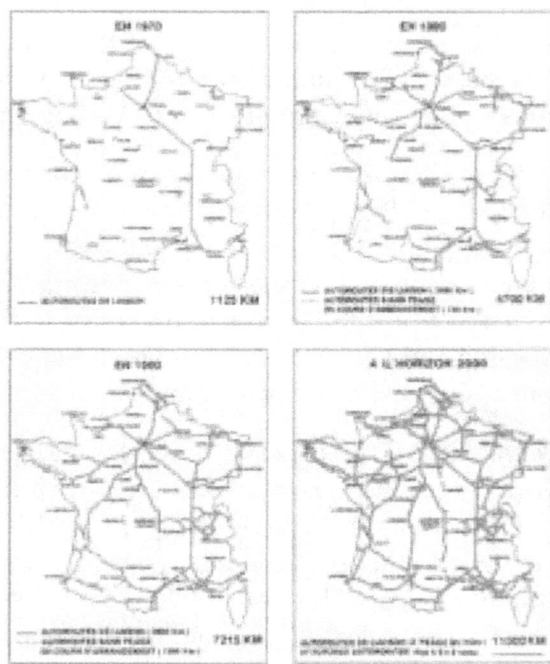

The nine concessionaires are listed in the table below. The adjacent figure graphically depicts the organization of the French government and the concessionaires.

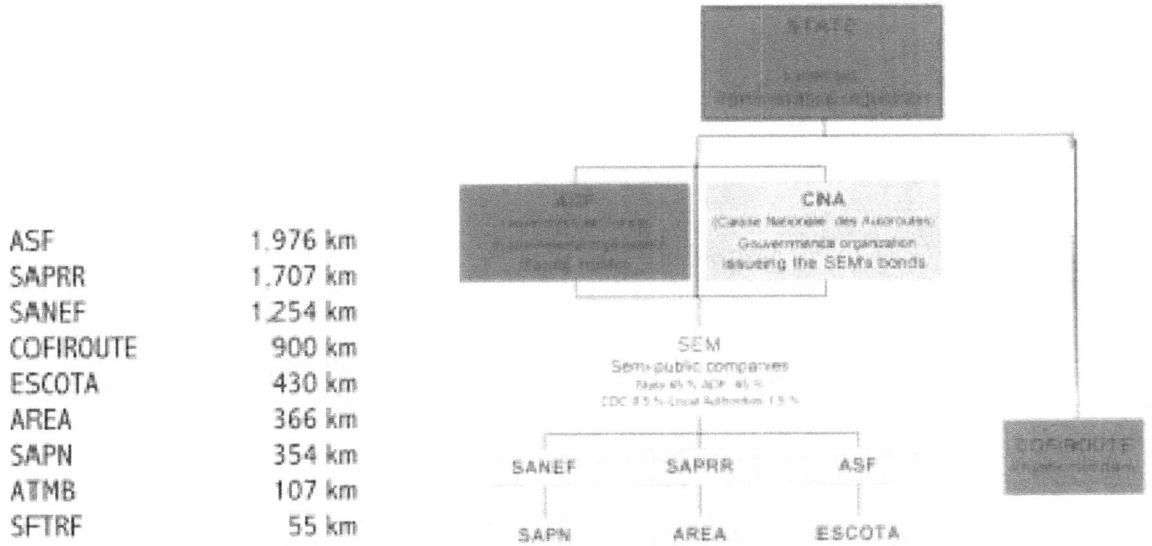

ASF	1,976 km
SAPRR	1,707 km
SANEF	1,254 km
COFIROUTE	900 km
ESCOTA	430 km
AREA	366 km
SAPN	354 km
ATMB	107 km
SFTRF	55 km

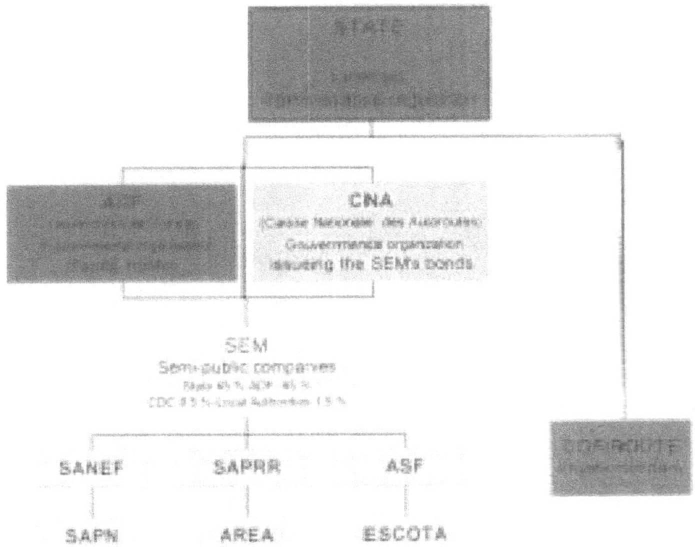

The relationship between the public owner and the concessionaire is very well defined. A 1985 law takes into account the client process, the quality, the cost, and the principles upon which the project is based. It also describes the roles of the engineer and contractor. The public sector client must state very clearly the needs of the public through a "program," which goes to the designers and contractors or the concessionaires. At the end of the project, the client approves the final product. The mission of the client is to define its needs in the "program" and assess the costs. This law is based on the principles that the client must participate throughout the entire process. When the public owner does not have the necessary expertise, it can employ an owner's representative or it can engage a firm to do the job in the name of the owner (but only in the construction phase). This responsibility cannot be totally transferred to third parties, as the law states that the owner must be present at all critical points in the process. As a result, these owners' representatives can only be public authorities or quasi-public agencies, and the owner must have a construction manager separate from the construction team. There are two milestones of cost assurance—one at the program level and one at the bidding time. If the bid costs are higher than the target, the engineer has to redesign.

French Concession Risk Allocation Strategy

Revenue and traffic risk	Concession
Construction cost risk	Concession
Financial risk	Concession
Operation cost risk	Concession
Project risk	French State
Force majeure	French State
Government action	French State

As with the Portuguese concession system, the government has carefully determined the appropriate risk allocation. The adjacent figure describes the risk allocation between the French government and the concessionaire. This table is based on the agreement between Cofiroute and the French government, but it is similar to agreements with the SEMs. The strategy is almost identical to that of the Portuguese government discussed in the previous section; however, the French government is willing to maintain more of the risk from new legislation. Although it is not shown in the table, Cofiroute and the other SEMs can purchase right of way on behalf of the government.

A description of Cofiroute's operations provides an excellent example of how the French concession program works. Cofiroute was formed in 1970 and has finished its first concession contract. It has been awarded numerous concessions through 2030 and, in fact, has one concession (the A86 West Loop) with a term ending in 2078 because the project is so large that a notice to proceed with the final segment will not be provided until one-half of the construction is done.

Cofiroute is in charge of designing and constructing the 900-km network. It does this through contracts with its contractor shareholders. It operates, maintains, and collects tolls on its network. Additionally, it is responsible for the safety and service of the customers. Cofiroute is contracted to keep the road available and safe; to restore normal traffic conditions in case of unforeseen events, including providing information to road users and public authorities; to operate traffic flows; to adjust demand to the actual capacity in order to limit or avoid congestion; to assist users; and to provide travel services. To maintain the network, Cofiroute provides emergency assistance through signing and coning, breakdown assistance, coordination with the police, employing emergency response plans, and implementing traffic management methods with other road operators.

It is interesting to note the diversified and central and regional government shareholding of the other eight quasi-public concessionaires. The central funding system is an efficient way of minimizing the cost of finance and of expanding the size of the network. However, both the French and EU authorities are seeking to make the French concession system for roads open to competition, as is already the case in other sectors such as water and wastewater treatment, in which the private companies play a substantial role.

U.S. Parallel: U.S. Highway 91 Express Lanes in California

Although not as prevalent as in Europe, concession contracts do exist in the United States. The U.S. Highway 91 express lanes in California provide a case study of how concessions can be successfully implemented. The following case study was written by Jean-François Poupinel, Chairman of Cofiroute, and Carl Williams, Deputy Secretary for Transportation, State of California (Perrot and Chatelus 2000).

> On July 29, 1989, as part of a package of bills that among other things would raise the state's gas tax by 9 cents a gallon over 4 years, the California legislature passed AB 680. Its drafters had proposed the legislation to test the efficacy of private involvement in public transportation facilities and to help compensate for the growing disparity between public resources and transportation needs. AB 680

authorized four infrastructure innovative demonstration projects using only private financing. The international competition solicited by Caltrans allowed prequalified private sector consortia to select any project that made both business and transportation sense. Out of thirteen international consortia that expressed interest, ten were prequalified. The "91 Express Lanes" was one of the four projects selected in September 1999. To date it is the only one that has been financed and constructed. [One other (the SR-125 project) is scheduled to close its financing during 2002.]

Located southeast of Los Angeles, California, the SR 91 is a very congested urban freeway linking three of the fastest growing Counties in the US: Riverside, San Bernardino and Orange Counties. Its eight lanes carry more than 250,000 vehicles per day, and is congested more than four hours a day in each direction. As a franchise holder, CPTC (California Private Transportation Company) has built four toll lanes (two in each direction) in the median of the existing non-tolled public highway. This new ten-mile-long facility operates without intermediate access points, and offers a fast, safe and reliable alternative to motorists who wish to save time.

CPTC is a limited partnership. The general partners are subsidiaries of Cofiroute Corporation (the U.S. subsidiary of Cofiroute) and of Peter Kiewit Sons who were in charge of the development and financing. They were joined by an affiliate of Granite Corporation Inc as a limited partner. The development (pre-construction) costs exceeded $ 10 million. The franchise is in force for 35 years. The major project risks were borne entirely by the concessionaire, who could elect to "abandon" the franchise without penalties if the project appeared infeasible. The Franchise Agreement precluded the use of any state or federal funds, but implemented innovative ideas concerning return on investment, performance incentives and the protection of the concessionaire through a non-competition zone.

This unique and innovative project has accumulated a number of "firsts". It is the first U.S. toll road to be privately financed in over 75 years. SR 91 is the world's first fully automated toll road, and is the first infrastructure project in the world to apply the concept of "value pricing". Since the "competition" offers a non-tolled ride a meter away, it continues to be important to listen at all times to the customers and to give them real "value" for money. To ensure that traffic flow remains fluid on the 91 Express Lanes now used by more than 30,000 vehicles per weekday, tolls were raised four times in the three-year life of the project. A major side benefit of the toll lanes is that traffic conditions have also been dramatically improved on the non-tolled public lanes.

This project, which has set the standard of a partnership between the private and the public sectors including the State of California (Caltrans) as well as local authorities, has received numerous awards including:

- The "Innovative Finance Award" by the Federal Highway Administration in 1992;
- The "Excellence in Transportation Award" by Caltrans in 1994;
- The "Innovative Project Award" by IBTTA in 1996.

The design and construction costs for the SR 91 project were financed through a taxable bank loan. An attempt by the concessionaire to restructure the deal, so as to allow a tax-exempt refinancing, proved controversial politically and was not implemented. The non-compete covenant has also proved to be politically controversial, and recently resulted in an agreement by the Orange County Transportation Authority to pay damages to the concessionaire in connection with construction of improvements to public facilities.

SELECTION OF CONCESSIONAIRES

As with selection of design-builders, selection of concessionaires can take many different forms. Add the fact that the government can in part own concessionaires, and the selection process becomes even more complicated. A purely qualifications-based selection had been employed by the French in the past, but they are turning to public competition for the selection of concessionaires today. This change is in part because of the new regulations of the EU, which is attempting to promote competition between EU countries. Refer to Chapter 3 for a summary of the EU policies for award of public works contracts, including concessions.

Two of the host countries, Portugal and the Netherlands, formally outlined their concessionaire process for the scan team. These processes are presented as two individual cases. However, as in the United States, selection systems vary on a project-by-project basis depending upon the characteristics of the projects and needs of the owner.

The Portuguese Concessionaire Selection Process

Because Portugal is operating under the new EU rules and is involved in a significant number of concession agreements, the Portuguese have created a rigorous and repeatable selection procedure. The legal framework for the selection was established and published for open procurement on the EU market. The procurement involves an international public tender in two stages, with no prequalification in the first phase.

The procurement process gives the concessionaires 5 months to prepare their proposals following receipt of the request for tenders. The proposals are then presented to the IEP. Since the proposals involve design, construction, operation, maintenance, and other services, the evaluation is quite complex and takes up to 6 months to complete. The proposals are evaluated on the following criteria:

- Technical quality;

- Government's financial effort;
- Expected net present value (NPV) of payments (SCUTs);
- Required subsidies;
- Level of risk and commitment;
- Proposed date for full operation; and
- Robustness of financial and contractual structure.

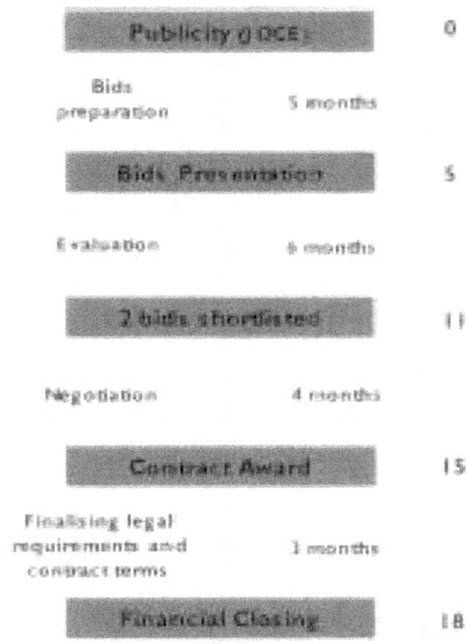

Two proposers are shortlisted after evaluation, and the IEP enters into competitive negotiations with the proposers. The final completion of the contract terms and contract award is accomplished in 3 months. The entire selection process takes an average of 18 months to complete.

The Dutch Concessionaire Selection Process and the Public-Private Comparator

The British and the Dutch are also giving careful consideration to the selection of concessionaires and public-private ventures. These countries have developed a tool for the evaluation of both the concession project and the concessionaire selection. The Public-Private Comparator (PPC) is employed to make a financial comparison of the viability of using a concession versus keeping a project in the private sector. The PPC compares the NPV of the concessionaire's proposal with the traditional cost of design, construction, maintenance, and operation in the traditional method. In this manner, the agencies can compare not only alternative concessionaires' proposals, but also the traditional procurement method.

The Dutch have incorporated the PPC into their selection procedure. The DBFM N31 road project provides an example of a selection process utilizing the PPC. The project involves improvements required for traffic safety/traffic flow reasons—the current road is one lane in each direction, and the new road will be two lanes in each direction. The road is approximately 25km in length, and includes one bridge and an aqueduct. The estimated construction cost is US$125 million. RWS determined the advantages and disadvantages of concession in a systematic and transparent way, using the PPC and based on an estimate of the NPV, cost, and benefits associated with the project. The study concluded that DBFM would be an efficient means of proceeding with the project, and also that there was no significant difference between DB and DBFM. Based on the PPC, RWS decided to proceed with the project as a DBFM pilot project. The proposed contract includes 15 years of maintenance. RWS also is asking for an alternative bid for 30 years of maintenance. The tender process is a combination of requirements of Dutch law and the EU Directive, and includes the following steps:

- **Advertisement:** Issuance of invitation to tender.

- **Pre-Selection:** Three to five selected – criteria for selection determined based on issues specific to project including experience with DBFM, experience with design, and construction of comparable projects.
- **Consultation (industry review):** RWS provides draft terms and conditions to contractors, holds an initial meeting with all prequalified contractors, holds one-on-one meetings with individual firms, receives written questions from contractors and provides written answers to all contractors.
- **Request for Bids:** Including final form of contract, allowing alternative bids associated with risk.
- **Bidding Phase:** Proposal includes design, risk analysis, draft quality plan, and financial proposal.
- **Negotiations:** Selection of two proposers for negotiations.
- **Best and Final Offers:** Selection based on "most economic bid" (best value) includes evaluation of design quality and the financial plan, including the NPV of payments to be made by RWS.

The contract will include incentive/disincentives based on road availability to encourage safety and minimize congestion. The contractor will be subject to penalty points for not following standard procedures to ensure safety. If too many penalty points are received, payment will be reduced. If the contractor's performance still fails to improve, at some point RWS will issue a warning followed by termination of the contract for cause. Please refer to Chapter 5: Performance Contracting – for more information on the Dutch incentive/disincentive system.

DURATION OF CONCESSIONS

Traditional U.S. contracts do not directly tie construction requirements to long-term performance. Once construction is complete, the contractor or design-builder typically provides a 1-year warranty. Concession agreements go far beyond simply warranting a project. By tying long-term operation and maintenance into the contract, the concept of warranty becomes irrelevant. Concessionaires are responsible for designing and constructing facilities that meet performance criteria over a long duration. This process creates a lifecycle mentality for the concessionaire from initial planning through contract closeout.

Durations of concessions were found to range from less than 5 years to more than 75 years, but the majority of concessions were contracted for 15 to 30 years. Maintenance contracts in the Netherlands and the United Kingdom tend to have a term in the range of 5 years. The majority of design-construction concessions of major motorways were in the range of 25 to 30 years. Many of the contracts also contained windows of profitability for determining the end of the contract because traffic forecasts for 30 years in the future are questionable. If traffic forecasts are wrong, there are only two options for equitable compensation for the project: change the rate of tolls (or payments) or change the duration of the contract. Political and financial viability typically limit changes in the rates charged. Possible solutions to problems caused by inaccurate traffic forecasts are to provide some mechanism for changing toll rates and, if necessary, changing the total duration of a concession to provide an equitable compensation to the concessionaire.

Another issue that must be addressed in concession agreements is the condition of the project upon delivery to the government at the end of the concession period. The Dutch are promoting concession periods with a duration equal to 75 percent of the design life of the product. This rule applies only when the concessionaire designs and builds the project. For maintenance and operation contracts on existing roads, the concessionaires in essence bid on the rights to operate and maintain the road in return for the toll collection or shadow toll payments from the government. The appropriate standard to be met at the end of contract life is not clear-cut in these situations. Concessions on existing facilities must be assessed on a project-by-project basis.

MEASURING THE PERFORMANCE OF CONCESSIONS

As described in the previous section, the role of a concessionaire goes far beyond simply warranting a project. Not only do the concessionaires have to maintain prescribed quality for the government, but they also must prove to their financial lenders and shareholders that they are delivering and maintaining a quality product. From what the host concessionaires described on the scan tour, these lenders and shareholders are sometimes more demanding than the States have ever been.

The question for the government is then how best to specify and measure the performance of the concessionaire. This question might best be answered through a discussion of the frameworks for the two concessionaires visited on the scan: the French concessionaire Cofiroute and the Portuguese concessionaire Autoestradas do Atlantico (AEA). Additionally, Chapter 5: Performance Contracting provides specific contract clauses for maintenance and operation contracts.

Given the nature of long-term contracts and high levels of competition for concessions globally, the concessionaires must maintain a high level of performance in order to remain competitive. It is well known that Cofiroute has one of the best asset-management systems in the world. In addition to its concessions in France, Cofiroute has concessions in the United Kingdom, South Africa, Los Angeles, Portugal, Argentina, Byelorussia, and others. It is the private ownership of the company that drives it to continuously measure the condition of and improve its assets—namely, its global highway network. The adjacent picture shows Cofiroute's pavement assessment vehicle, which is used to continuously monitor the condition of its roadways. Cofiroute boasts one of the most sophisticated and technically advanced asset-management systems of any private company or public agency.

AEA has a shorter history than Cofiroute, but its function is essentially the same. AEA is located north of Lisbon in a rural but growing area. AEA runs a concession on a highway that is about 8 to 10 years old. It believes that the traffic

will increase substantially as Lisbon grows. The performance terms of AEA's contract includes:

- The Concessionaire must keep Motorways in very good conservation and perfect condition of utilization, carrying out all the necessary works in order to permanently satisfy the Motorways purposes.
- The Concessionaire is responsible for the high standards of conservation and functioning of environmental monitoring equipment, environmental conservation and preservation systems and noise protection system.
- The Concessionaire must respect minimum quality standards, such as pavement bond and smoothness, conservation of signaling, clients assistance and safety equipment.
- Specific performance tests include:
  - Tests with FWD every 100m, including visual inspections
  - Longitudinal irregularities determination
  - Pavement depression due to heavy traffic measures
  - Friction measures
  - Pavement degradation report
- It has four separate performance contracts:
  - Contract 1:
    - Vegetation (shrubs and plants) maintenance
    - Cleaning and sweeping
    - Fencing, repairing and maintenance
  - Contract 2
    - Safety equipment, repairing and conservation
    - Traffic signs, road signs and safety guards
  - Contract 3
    - Civil engineering works, conservation and maintenance
    - Drainage
    - Sloping banks
    - Pavements
    - Concrete structures
  - Contract 4
    - Engineering structures, maintenance

Regardless of whether the agencies are responsible for measuring concession performance or the concessions measure their own performance, with performance audits undertaken by the agencies, performance measurement and benchmarking are the cornerstones for the success of any concession contract. More specific information concerning the measurement and assessment of long-term contracts is provided in Chapter 5: Performance Contracting.

SUMMARY

Motorways in Europe utilize concessions for both construction and maintenance. The long-term nature and best-value selection of a concessionaire provides the opportunity to benchmark and achieve exceptional performance. All of the host countries visited on this contract administration scan are incorporating concessions into their strategic networks plans—some at a smaller maintenance and operation level and some for the majority of their networks. Appropriate selection of concession projects and concessionaires will be one of the keys to successful incorporation of this contracting strategy into the United States. Agencies must consider the specifications and measurement of performance criteria carefully because typical concessions last for 25 to 30 years. Used in appropriate circumstances, concession contracts may prove to be very beneficial to the U.S. highway sector. The scan team recommends that the following issues be explored in the United States as a means to speed the delivery of our infrastructure and increase the quality of construction and maintenance:

- Investigate and document the long-term performance of concessions globally as a benchmark for current U.S. finance, design, construction, operations, and maintenance procedures.
- Consider more closely the use of concessions on existing networks and new facilities when creating U.S. strategic network plans.
- Develop policies that allocate risks to the parties who can best manage them in concession contracts, specifically in the areas of environmental and right-of-way issues.
- Employ a Public-Private Comparator that is appropriate for use in the United States when making the economic decisions surrounding concession contracts.
- Develop appropriate performance specifications and measurement standards to routinely apply on concession contracts to ensure consistency from project to project and State to State.

Chapter 8: Recommendations

The 2001 contract administration scan team was privileged to be able to travel to Europe and visit with the representatives of five host countries (France, the Netherlands, Portugal, Sweden, and the United Kingdom). The team witnessed many alternative contract administration practices and unanimously believes that a number of these can be immediately applied in the United States.

Following the European tour, the scan study moved into the scanning technology implementation phase (STIP). All team members are actively implementing practices that are applicable to their positions in the transportation industry, but a smaller STIP team was created to move the implementation forward quickly. The STIP team met to determine the most appropriate concepts for implementation from the numerous innovations and best practices discovered on the scan.

The STIP team developed a questionnaire concerning the top 13 concepts likely to have the most potential to impact the U.S. highway industry. All of the scan team members responded to the questionnaire. The entire scan team rated the implementation options on a scale from "5 – Extremely Important" (idea or recommendation is very critical and will significantly improve contract administration procedures or project delivery methods) to "1 – Not Important" (idea or recommendation will not significantly improve current practices or solve any real problem). The table below provides the ranking for the implementation items.

Rank	Category	Implementation Concept Description
1	Contracting Techniques	Best Value Procurement Techniques
2	Contracting Techniques	Procurement Utilizing Confidential Negotiation Processes
3	Performance Contracting	Performance Specifications
4	Performance Contracting	Long-Term Maintenance Contracts
5	Performance Contracting	Quality Control is Sole Responsibility of Contractor (penalty card quality systems, etc.)
6	Design-Build	Lifecycle Cost Award
7	Alternative Financing	Opportunities for Joint Development
8	Contracting Techniques	Alternative Payment Mechanisms (user-based payment mechanisms, e.g., product availability, milestone pay points, contractor determines pay quantity, etc.)
9	Concessions	Performance Metrics for Concessionaire Selection and Payment
10	Contracting Techniques	Alternative Contract Types (framework, managing agency contracting, integrated supply chain management, etc.)
11	Concessions	Integrating Concessions into Long-Term Planning
12	Alternative Financing	Use of Shadow Tolls (either based upon use or performance)
13	Asset Management	Valuation of Assets

All 13 of the implementation concepts are important, but the top three were determined to be the focus for implementation because of their high potential impact and relative ease of implementation. These three techniques, and six of the top eight, are found in the contracting techniques and performance contracting categories. These three concepts are presented as primary findings below. The other concepts are summarized as additional findings in the order in which they are found in this report.

PRIMARY RECOMMENDATIONS

The United States has much to learn from European highways agencies. In particular, the best-value approach combined with the ability to negotiate technical terms and alternative concepts with selected contractors has enabled European agencies to award contracts at reasonable cost to those providers with a proven track record for responsiveness to the public sector's needs. The FHWA and the State DOTs should consider more use of best-value negotiated contracts, giving contractors the opportunity to develop reputations that enable them to be exceptional performers and compete in best-value procurements. The FHWA and the DOTs also should consider moving toward the use of more performance specifications, which will allow the private-sector industry to innovate and continuously improve the quality, efficiency, and safety of the highway system. These performance specifications will require associated performance indicators to measure and benchmark exceptional performance. Specifically, the scan team recommends that the following concepts and tools be explored in the United States as a means to speed the delivery of our infrastructure and increase the quality of construction and maintenance.

- Use best-value award techniques in the selection wherever it is shown that value can be added through quality or innovation.
- Explore techniques to fairly and equitably employ confidential negotiations and discussions of alternative proposals to capitalize on the creativity and innovation of the private sector.
- Create consistent performance specifications that define the owner's performance objectives, which can be used to promote consistency in specifications while allowing for innovation in design, construction, and maintenance.
- In conjunction with the performance specification system, develop consistent and objective performance indicators that allow for the measurement and verifiable benchmarking of the performance specifications. These performance indicators should be used to create a system of continuous improvement for the industry.
- Explore the formation of an audit group, similar to the U.K. PRIDe group, as a means to benchmark performance indicators for use by all States. This team will be able to ensure, through diligent benchmarking, that projects are being delivered at competitive costs in lieu of ensuring competitive costs through our current low-bid system.

ADDITIONAL RECOMMENDATIONS

The primary recommendations above will be the focus of the STIP team. However, the scan tour revealed other concepts that have the potential to improve the U.S. highway industry. The scan team recommends that efforts be made to employ these concepts whenever opportunities arise. These additional concepts are ranked in the table above and presented here in the order in which they are found in this report.

Contracting Techniques

The European highway community is benefiting from widespread use of best-value procurement, greater latitude to enter into competitive negotiations, more use of alternative designs in proposals, extensive use of management contracts, long-term contracts tying maintenance to construction, and payment methods that are based on outcomes at the end of the projects rather than payment for work as it is put in place. In addition to the primary contracting techniques listed in the primary recommendations above, the scan team recommends that the following concepts be implemented:

- Use management contracting on repetitive work where project characteristics display a potential to save construction and procurement costs.
- Explore integrated supply chain contracts to capitalize on the efficiencies documented in the manufacturing sectors.
- Test payments by milestones and payments by availability as a way to tie quality performance to payment structures.

Design-Build

The European countries visited on the scan tour provided the team with many valuable design-build contracting insights. The primary lessons learned on this scan tour relate to the types of projects utilizing design-build, the use of best-value selection, the percentage of design in the solicitation, design and construction administration, third-party risks, the use of warranties, and the addition of maintenance and operation to design-build contracts. The scan team recommends that the following concepts be implemented:

- Capitalize on best-value selection processes to promote competition and innovation among design-builders.
- Promote appropriate use of performance specifications with low levels of design in design-build RFPs to promote innovation and accountability from the private-sector proposers.
- Assign third-party risks to the party in the contract that can best control them.
- Ensure construction quality and cultivate a pool of qualified lifecycle service providers through the incorporation of maintenance and operation into design-build projects.

Performance Contracting

The scan team discovered applications of performance contracts in Europe for long-term maintenance, DBM, and concession contracts. The essential lessons learned on performance specifications on this contract administration scan can be summarized into the categories of performance specifications, performance indicators, warranties, and QA/QC. The scan team recommends that the following concepts be implemented:

- Catalog those performance contracting methods currently in use in the U.S. transportation industry.
- Employ an aggressive pilot study program exploring the use of performance contracting for both construction and maintenance to determine the efficiency of current methods and to develop consistent and objective performance indicators.

This will allow for the measurement and verifiable benchmarking of the performance and a trial of other promising performance contracting methods.

- Nationally benchmark the performance of long-term warranties against the use of performance contracts to determine which system provides better value to the public.
- Promote the U.S. trend for contractor-controlled quality control programs and develop incentive/disincentive systems for quality such as the red card/yellow card system used in Europe.

Alternative Finance

European highways agencies are working closely with private-sector partners to finance, build, and maintain projects that are not viable using traditional funding mechanisms, either because of lack of funding or sociopolitical constraints. These alternative funding sources take a whole-life approach to project design, construction, and maintenance. Alternative payment methods offer solutions that increase price competition from the private sector, but also incentivize and improve quality in the completed and maintained project. The scan team recommends that the following concepts be implemented:

- Take a whole-life approach to planning through linking construction quality and maintenance to private financing.
- Leverage innovative concepts from the private sector to overcome social and political barriers for improving the infrastructure.
- Explore the use of shadow tolls and AMPMs as a means to defer infrastructure payments while decreasing congestion and increasing safety and availability.
- Measure and benchmark the performance of these alternative payment mechanisms to create an opportunity for continuous improvement measurements.

Concessions

Concessions are commonly used for both construction and maintenance of European motorways. The long-term nature and best-value selection of a concessionaire provides the opportunity to benchmark and achieve exceptional performance. Used in appropriate circumstances, concession contracts may prove to be very beneficial to the U.S. highway sector. The scan team recommends that the following concepts be implemented:

- Investigate and document the long-term performance of concessions globally as a benchmark for current U.S. finance, design, construction, operations, and maintenance procedures.
- Consider more closely the use of concessions on existing networks and new facilities when creating U.S. strategic network plans.
- Develop policies that allocate risks to the parties that can best manage them in concession contracts, specifically in the areas of environmental and right-of-way issues.

- Employ a public-private comparator that is appropriate for use in the United States when making the economic decisions surrounding concession contracts.
- Develop appropriate performance specifications and measurement standards to routinely apply on concession contracts to ensure consistency from project to project and State to State.

Bibliography

British Highways Agency. (2000). *Active Management Payment Mechanism For Future DBFO Projects: Summary of Principles.* British Highways Agency, London, England.

British Highways Agency. (1999). *An Introduction to The Highways Agency.* British Highways Agency, London, England.

British Highways Agency. (1997). *DBFO – Value in Roads.* British Highways Agency, London, England.

British Highways Agency. (1999). Framework Document. British Highways Agency, London, England.

British Highways Agency. (2001). "Procurement Guidance Strategy Note," Framework Contract, Issue No. 2, Revision No. 0. British Highways Agency, London, England.

British Highways Agency. (2001). "Performance Indicators, Annex 12," Model Document MAC, Issue No. 3, Revision No. 1. British Highways Agency, London, England.

Burns, J. (December 1999). Paving the Way, A Consultation Paper: A Review of the Management and Maintenance Arrangements for Motorways and Trunk Roads in England. British Highways Agency, London, England.

Connaughton, J.N. and Green, S.D. (1996). *Value Management in Construction: A Client's Guide.* Construction Industry Research and Information Association Special Publication 129. London, England.

Jackson-Robbins, A. (1998). *Selecting Contractors by Value.* Construction Industry Research and Information Association Special Publication SP 150. London, England.

Perrot, J.-Y. and Chatelus, G., eds. (2000). *Financing of Major Infrastructure and Public Service Projects: Public-Private Partnership, Lessons from French Experience Throughout the World.* French Ministry of Public Works, Transport and Housing, Economic and International Affairs Division, Paris, France.

MES Intervenção Operacioinal dos Transportes. (1999). *Public Private Partnerships.* MES Intervenção Operacioinal dos Transportes, Lisbon, Portugal.

Richmond-Coggan, D. (2001). *Construction Contract Incentive Scheme – Lessons from Experience.* Construction Industry Research and Information Association C554 London, England.

Santana, F. and Calado, R.H., Gattel. (1999). *Ponte Vasco da Gama Bridge.* Ministerio do Equipamento, do Planeamento e da Administracao do Territorio, Secretaria de Estado das Obras Publicas, Lisbon, Portugal.

Appendix A: Scanning Team Members

Team Member Affiliations
(at time of scan tour)

David O. Cox (Co-Chair)
Division Administrator
Federal Highway Administration—
Oregon Division
530 Center Street, N.E., Suite 100
Salem, OR 97301-4137
Tel: (503) 399-5749
Fax: (503) 399-5838
Email: david.cox@fhwa.dot.gov

Ronald C. Williams (Co-Chair)
State Construction Engineer
Arizona Department of Transportation
206 South 17th Avenue (MD 172-A)
Phoenix, AZ 85007
Tel: (602) 712-7323
Fax: (602) 254-5128
Email: rwilliams@dot.state.az.us

Keith R. Molenaar (Report Facilitator)
Assistant Professor
Department of Civil, Environmental & Architectural Engineering
University of Colorado at Boulder
Campus Box 428, ECOT 411
Boulder, CO 80309-0428
Tel: (303) 735-4276
Fax: (303) 492-7317
Email: keith.molenaar@colorado.edu

James J. Ernzen
Associate Professor
College of Engineering and Applied Sciences
Del E. Webb School of Construction
Arizona State University
Box 870204
Tempe, AZ 85287-0204
Tel: (480) 965-0389
Fax: (480) 965-1769
Email: james.ernzen@asu.edu

Charlie Franklin (Frank) Gee
Construction Engineer
Virginia Department of Transportation
1401 East Broad Street
Richmond, VA 23219
Tel: (804) 786-2783
Fax: (804) 786-7778
Email: gee_cf@vdot.state.va.us

Gregory G. Henk (Representing ARTBA)
Senior Vice President
HBG Constructors Inc.
31441 Santa Margarita Pkwy
Suite A-160
Rancho Santa Margarita, CA 92688
Tel: (949) 589-3343
Fax: (949) 709-2639
Email: ghenk@flatironstructures.com

Jeff W. Kolb
Supervisory Highway Engineer
Federal Highway Administration,
California Division
980 9th Street (Suite 400)
Sacramento, CA 95814-2724
Tel: (916) 498-5037
Fax: (916) 498-5008
Email: jeff.kolb@fhwa.dot.gov

Tanya C. Matthews, AIC (Representing Design-Build)
Senior Vice President
Director Design-Build Development
Parsons Brinckerhoff Constructors, Inc.
465 Spring Park Place
Herndon, VA 20170-5227
Tel: (703) 742-5721
Fax: (703) 742-5962
Email: matthewst@pbworld.com

Len Sanderson
State Highway Administrator
North Carolina Department of Transportation
1536 Mail Service Center
Raleigh, NC 27699-1536
Tel: (919) 733-7384
Fax: (919) 733-9428
Email: lsanderson@dot.state.nc.us

Nancy C. Smith
Partner
Nossaman, Guthner, Knox & Elliott
445 South Figueroa Street (31st Floor)
Los Angeles, CA 90071-1602
Tel: (213) 612-7837
Fax: (213) 612-7801
Email: nsmith@nossaman.com or ncs@ngke.com

Gary C. Whited
Deputy Administrator
Wisconsin Department of Transportation
4802 Sheboygan Avenue (Room 451)
P.O. Box 7965
Madison, WI 53707-7965
Tel: (608) 267-7774
Fax: (608) 264-6667
Email: gary.whited@dot.state.wi.us

John W. Wight (Representing ARTBA)
Executive Vice President
HNTB Corporation
Wayne Plaza I, Suite 400
145 Route 46 West
Wayne, NJ 07470
Tel: (973) 237-3006
Fax: (973) 237-1673
Email: jwight@hntb.com

Gerald Yakowenko
Contract Administration Engineer
Federal Highway Administration
Contract Administration Group
Office of Program Administration, HIPA-30, Room 3134
400 Seventh Street, S.W.
Washington, DC 20590
Tel: (202) 366-1562
Fax: (202) 366-3988
Email: gerald.yakowenko@fhwa.dot.gov

Team Member Biographic Sketches
(at the time of the scan tour)

David O. Cox (Panel Co-Chair) is the Federal Highway Division Administrator for the State of Oregon. In this capacity he has overall responsibility for the Federal Aid Highway System and the expenditure of Federal-aid Highway Funds in the State. Mr. Cox has more than 30 years of experience with the FHWA. He is a registered professional engineer and holds BS and MS degrees in Civil Engineering. In previous assignments, Mr. Cox led the Contract Administration Office in Washington, D.C., where he vigorously promoted alternative contracting techniques throughout the nation. He is one of the principle authors of the soon to be released "FHWA Design / Build Regulations," and led the team that wrote FHWA's "Financial Plan Guidance" for projects with costs exceeding $1 billion. Mr. Cox has held the national positions of Secretary to the AASHTO Subcommittee on Construction and Chairman of the American Society of Civil Engineers Subcommittee on Highway Construction and Maintenance.

Ronald C. (Ron) Williams (Panel Co-Chair) is the State Construction Engineer, Headquarters Office, Arizona DOT in Phoenix, and is responsible for contract administration policies, procedures, and specifications. He has 32 years of experience with the Arizona DOT, including as Resident Engineer, Area Engineer, and Assistant District Engineer. As State Construction Engineer, Mr. Williams is responsible for implementation of alternative contracting practices and management of the design-build process, with contracts to date for design-build projects of $380 million dollars. Mr. Williams is the Chairman of the Contract Administration Task Force, AASHTO Subcommittee on Construction and chairs the AASHTO Design/Build Rules Committee. He served on the interdisciplinary team that helped the Arizona State Legislature develop new Alternative Contracting Methods legislation. He is the Industry Co-Chairman on the Alternative Contracting Methods Task Force, Alliance for Construction Excellence at Arizona State University. Mr. Williams holds a BS degree in Civil Engineering from the University of Arizona and is a registered professional engineer in the State of Arizona. He is a member of the American Public Works Association and the American Society of Civil Engineers (ASCE).

Dr. Keith R. Molenaar (Report Facilitator) is an Assistant Professor with the Construction Engineering and Management (CEM) Program in the Department of Civil, Environmental and Architectural Engineering at the University of Colorado at Boulder. His research focuses on alternative delivery strategies for the procurement of infrastructure and constructed facilities. His responsibilities include the coordination of a collaborative research effort aimed at exploring alternative delivery methods, analyzing project performance, and disseminating research results to owners, designers, constructors, and students. Dr. Molenaar was previously a faculty member at the Georgia Institute of Technology were he was Group Leader of the Construction Research Center's Procurement and Project Delivery research initiative. Dr. Molenaar has a BS degree in Architectural Engineering and MS and PhD degrees in Civil Engineering from the University of Colorado at Boulder. Dr. Molenaar is an active member of ASCE, the Design-Build Institute of America (DBIA), and the Construction Management Association of America (CMAA).

Dr. James J. Ernzen is an associate professor at the Del E. Webb School of Construction at Arizona State University. He currently serves as the Co-chairperson of the Alternative Project Delivery Methods Taskforce at the university where he teaches a graduate course on Design-Build Project Delivery Methods and conducts research in the area of design-build and other alternative project delivery systems. Prior to joining the Construction School, he spent more than 21 years as a construction manager, project engineer, construction materials researcher, and civil engineering educator in the Army Corps of Engineers. Dr. Ernzen holds BS and MS degrees from the University of Notre Dame and a Ph.D. from the University of Texas at Austin. He is a licensed professional engineer in the Commonwealth of Virginia and serves on both the Education and Civil/Infrastructure Committees for the DBIA and the Construction Management Committee for the TRB.

Charlie Franklin (Frank) Gee is currently the State Construction Engineer for the Virginia Department of Transportation. He is the current Vice Chairman of the AASHTO Subcommittee on Construction, which is one of the sponsoring entities of this scanning tour. He is interested in evaluating the best business techniques in developing and administering transportation contracts so all members of AASHTO can benefit. In positions as Resident and Construction Engineer, Mr. Gee has been involved in construction and the administration of contracts for more than 25 years. The current program in Virginia does approximately 700 contracts at a value of over $1 billion. Mr. Gee is a civil engineering graduate of North Carolina State University and is affiliated with several organizations.

Gregory G. Henk is Executive Vice President of Flatiron Structures' Design/Build Division and is responsible for project development and management. He has 30 years of project management experience in urban transportation and planning, engineering design, and construction. Mr. Henk recently served as Executive Vice President of the Transportation Corridor Agencies in Orange County, California, where he oversaw the design, construction, and toll operation of three toll roads with a combined construction value in excess of $2 billion. He has worked for State DOTs, public transportation agencies, and private engineering and construction firms. A recognized leader in the design/build industry, Mr. Henk is a specialist in public and private highway financing activities. He has been involved in the issuance of $2.8 billion of non-recourse, startup toll revenue bonds.

Jeff W. Kolb is the Supervisory Transportation Engineer for the South Program Delivery Team in the California Division of the FHWA in Sacramento, California. He is responsible for federal oversight on all federally funded highway transportation projects in the southern half of California. This includes several "mega-projects" (costing over $1 billion) at least one of which is a design-build project. Mr. Kolb has worked with the FHWA for more than 17 years, with previous assignments in the Utah Division, the Florida Division, and the Regional Office in Denver, Colorado. His experiences on these assignments included an integral role on the $1.59 billion Interstate 15 Design/Build Reconstruction Project in Salt Lake City, Utah; early design-build activities and alternative contracting in the Florida Division; and experience in the innovative arena of Intelligent Transportation Systems in the Regional Office assignment. Mr. Kolb is a graduate of North Carolina State University with a BS in Civil Engineering. He also holds a master's degree in Civil Engineering from Florida State University. He is a licensed professional engineer in the State of Florida.

Tanya C. Matthews, AIC, is the Vice President, Government Affairs for the Design-Build Institute of America. She is responsible for changes in federal law for design-build and she works to broaden the use of design-build in federal agencies in America. Prior to joining DBIA full time, Tanya was the Senior Vice President and Director of Design-Build Development for Parsons Brinckerhoff (PB), one of the world's largest engineering firms, and she was responsible for the expanded development of PB's design/build practice. A recognized leader in design-build, she served as 1999 Chairman of the Board of the DBIA and has played a decisive role in facilitating greater acceptance of the design-build process with public agencies and private-sector owners in the 50 States and abroad. Under Ms. Matthews' leadership, the share of design-build in the domestic, nonresidential market grew from 27 percent to 32 percent, and she was instrumental in gaining congressional approval for design-build on federal projects. Ms. Matthews holds five academic degrees, including an MBA from the University of Maryland, and two BS degrees from The American University.

In addition to her affiliation with the DBIA, Ms. Matthews is a member and past President, Chairman of the Board, and Chairman of the Executive Committee of the Dulles Area Transportation Association (DATA). She also has served as President of the Metro-Washington Chapter of the American Institute of Constructors and was the Founding President of their Northern Virginia Chapter. She won the "Marketer of the Year" award from the Society for Marketing Professional Services, and has been Awards Committee Chairman for the Northern Virginia Chapter of the Associated Builders and Contractors since 1998.

Len Sanderson is the State Highway Administrator for the North Carolina DOT. Mr. Sanderson currently directs the planning, design, construction, and maintenance of the 78,500-mile State highway system in North Carolina. His areas of emphasis are work zone safety and dissemination of information to motorists. Prior to his appointment as Highway Administrator, he served as the Construction Branch Manager for statewide operations. Mr. Sanderson is a graduate of North Carolina State University and holds a bachelor's degree in Civil Engineering. He is a licensed professional engineer and serves on several AASHTO committees.

Nancy C. Smith is a partner with the law firm of Nossaman, Guthner, Knox & Elliott, LLP in Los Angeles, California. Her area of specialty is construction and procurement law, particularly design-build issues and public-private partnerships. Ms. Smith has been instrumental in developing some of the largest transportation infrastructure projects in the United States, including public toll roads in Orange County, California (design-build contracts totaling over $2 billion), the Alameda Transportation Corridor in Los Angeles (a $712 million design-build contract), the I-15 Reconstruction Project in Salt Lake City, Utah (a $1.4 billion design-build-maintain contract), the Hudson-Bergen Light Rail Transit Project in New Jersey (an $800 million design-build-operate-maintain contract) and the T-Rex I-25 corridor reconstruction in Colorado (an $1. 8 billion design-build project being jointly procured by the Colorado DOT and the Denver Regional Transit District. She holds a BA in German from the University of Florida (with highest honors, Phi Beta Kappa) and a JD from Yale Law School. She is a member of the Women's Transportation Seminar (Board member, Los Angeles Area Chapter), the DBIA (Board member, Western Pacific Chapter), and the TRB's Public Contract Law Committee.

Frederick Werner is an Innovative Finance Specialist for the FHWA at the Southern Resource Center in Atlanta, Georgia. Mr. Werner currently provides technical assistance to States, local governments, and private-sector firms in the use of federal financing techniques to develop innovative or alternative financing for transportation projects. Among the federal financing techniques currently promoted are subordinate federal credit and State bonding secured by future federal funds. Prior to joining the Southern Resource Center in 1999, Mr. Werner served as the Financial Manager in the FHWA Puerto Rico Division Office. He is a graduate of Marquette University (Milwaukee, Wisconsin) with bachelor's degrees in Accounting and Finance. Mr. Werner holds CPA certification in Wisconsin and Illinois.

Gary C. Whited is the Deputy Administrator of the Division of Infrastructure Development for the Wisconsin DOT (WisDOT) in Madison, Wisconsin. As Deputy Administrator, Mr. Whited shares responsibility with the Division Administrator for managing the headquarters engineering division of WisDOT that is responsible for the development and operation of Wisconsin's highways, airports, harbors, and railroads. The division focuses on Statewide development of policies and standards for these transportation systems, which includes management of construction programs. He has been with WisDOT for 29 years, 10 years of which he was Director of the Bureau of Highway Construction where he was directly responsible for administering the Statewide highway construction program. Mr. Whited holds a BS degree in Civil Engineering from Iowa State University and an MS degree in Civil Engineering from the University of Wisconsin-Madison. He is a member of the AASHTO Subcommittee on Construction and currently serves on the Advisory Committee for the Construction Engineering and Management Program at the University of Wisconsin- Madison. He also serves on numerous TRB, AASHTO, and NCHRP committees, is currently the Director of the Midwest Concrete Consortium, and has authored papers published in *ASTM Geotechnical Testing Journal*, *Transportation Research Record*, and *Proceedings of the International Conference on Geotechnical Engineering*. He is a registered professional engineer in the State of Wisconsin and is a past ASCE Madison Branch president and past member of the Wisconsin ASCE Section Board of Direction.

John W. Wight is an Executive Vice President with HNTB Corporation, a large national A/E firm specializing in transportation. He also is Chairman of the ARTBA. ARTBA is the nation's oldest (established 1902) national construction association and is a recognized leader in the transportation construction industry. It is the only association that exclusively represents the collective interests of all sectors of the transportation construction industry before the White House, Congress, federal agencies, media, and the public. In recent years, issues of alternative contracting methods have been at the forefront of ARTBA's concerns, especially as they relate to pending federal legislative changes. Throughout his 33-year career with HNTB, Mr. Wight has been involved in all aspects of highway design and administration. He has worked extensively on toll highways across the United States and has been deeply involved in the transportation privatization movement for the past 10 years. Mr. Wight has his bachelor and master's degrees in Civil Engineering from Cornell University. He is a licensed professional engineer in 18 states and is a member of ASCE, NSPE and IBTTA in addition to his involvement in ARTBA.

Gerald Yakowenko is a Contract Administration Engineer with the FHWA in Washington, D.C. Mr. Yakowenko currently is responsible for developing and interpreting FHWA's policy for federally funded construction contracts. In addition, he provides technical and programmatic assistance to the State DOTs for innovative contracting practices such as design-build and multi-parameter bidding contracts. Throughout his 21-year career with FHWA, he has been involved with construction contracting issues. He is a 1977 graduate of Lehigh University and holds a BS degree in Civil Engineering. He is a licensed professional engineer in the State of Missouri. Mr. Yakowenko is a participating member of the ASCE, the AASHTO Subcommittee on Construction, and the TRB Panel A2F05 – Committee on Construction Management.

Appendix B: Contacts in Host Countries

Listed below are the names of the individuals with whom the scan team met during the trip. The contract administration scan team wishes to express its sincere gratitude to these individuals for their generous donation of time and the valuable information that they provided to the team. The listings are presented in alphabetical order.

UNITED KINGDOM

British Highways Agency

Mr. Dave Ball
Mr. Bob Gale
Mr. Les Hawkero
Mr. Lionel Jones
Mr. Yogesh Patel
Mr. Stephen Phillips
Mr. John Powell
Mr. Steve Rowsell
Mr. Graham Taylor

Carillion

Mr. Keith Fountain
Dr. Peter Forsyth
Mr. Tony Gates
Ms. Sue Tester
Mr. Paul Woodman

Construction Industry Research and Information Association (CIRIA)

Mr. Peter Bransby
Mr. David Churcher
Mr. Keith Montague

Halcrow Transportation Infrastructure

Mr. Philip Alexander (Halcrows)

URS – Dames & Moore, O'Brien Kreitzberg, Thorburn Colquhoun

Mr. Chris Darling
Mr. Keith Holloway
Mr. David S. Weeks

FRANCE

Ministry for Infrastructure, Transport and Housing

Ms. Nolwenn David-Nozay
Ms. Yolande Daniel
Mr. Philippe Gratadour
Mr. Philippe Leger

Ministere De L'Equipement, Des Transports Et Du Tourisme

Ms. Nicole Sitruk

Service Grandes Infrastructures

Ms. Estelle Brachlianoff

Cofiroute

Mr. Dominique Ratouis

THE NETHERLANDS

Ministry of Transport, Public Works and Water Management

Mr. Bert van Andel
Mr. Herman Gerrits
Mr. J. Frits Houtman
Mr. P. Keift
Mr. Arie L. Korteweg
Mr. Paul van der Kroon
Mr. Wim L. Leendertse
Mr. A.P.H. (Toine) van Liebergen
Mr. Charles H.N.M. Petit
Mr. Jan Swart

ARCADIS BOUW/INFRA BV

Mr. Rolf A. Muller

THE NETHERLANDS

Delft University of Technology

Prof. dr. Ir. Hennes A. J. de Ridder

Grontmij Data & Infrastructuur

Mr. Paul J.A. Oortwijn

Hollandsche Beton Groep nv

Mr. F. (Fedde) Tolman

IBASE bv

Mr. Nicholas Nass

N.V. Westerscheldetunnel Co-operation

Mr. Jaap Heujboer

Strukton, Betonbouw

Mr. Hans Wenkenbach

Sytwende, Voorburg

Mr. Stan Rostenberg

Tunnel Groene Hart

Mr. Wim L Leendertse

Volker Wessels Stevin

Mr. F.J.M. van den Bergh

PORTUGAL

Instituto das Estradas de Portugal

Ms. Carla Barradas
Mr. Herculano Dos Santos E Sousa
Mr. Pedro M.M. Santana
Mr. Rui Silva Oliveira
Mr. Rui Neves Soares
Ms. Lidia Soares Nobre
Ms. Rita Susana da Cruz Serrafo
Mr. Jorge Zuniga de Almeida Santo

Auto-Estradas Do Atlântico

Mr. José Barreto
Ms. D. Ana Rita Bednell
Mr. João Ceia,
Mr. Leonardo Cruz
Ms. Ana Jerónimo
Ms. Ana Magalhães
Dr. Silva Pereira
Mr. Câmara Pestana
Mr. Pedro M.M. Santana

Consulpav

Dr. Jorge Barreira de Sousa

BPI

Dr. Miguel Morais Alves
Ms. Dra. Maria João Cabral

ACESA

Mr. D. Pere Rigau

SWEDEN

National Road Management Department

Mr. Gunnar Tunkrans
Mr. Hakan Johnson

Appendix C: Amplifying Questions

CONTRACT ADMINISTRATION PANEL

The following list of questions contains six (6) topics that the U.S. panel would like to discuss with you. These questions are intended to clarify and expand upon the Panel Topics of Interest described in the Panel Overview paper. Each of these six topics is subdivided into more specific categories. Use these questions as a guide, but please comment on any innovative techniques, particularly successful experiences or not-so-successful experiences, that our panel has overlooked in the questions.

I. Contracting Techniques

General Definitions

I-1. What is the "traditional" method of construction contracting in your country? In the United States, we consider the design-bid-build method to be our traditional system.

I-2. What other contracting techniques is your country currently using or considering?

I-3. Were there any governmental or legislative issues that inhibited the use of these techniques? If so, how were they overcome?

Project and Delivery Team Selection

I-4. If your country uses multiple contracting techniques, what are the issues of concern or criteria considered when selecting a particular contracting technique for a project?

I-5. What parameters, in addition to cost, does your country use to select contractors or teams? For example in the United States, we use both "low-bid," which is based solely on price, and "best-value," which is a combination of price and technical factors. Are there other methods for contractor or team selection?

I-6. How are design firms selected when competing on innovative design projects?

I-7. How are construction contractors selected to maximize innovation and/or early completion?

Global Effects

I-8. What have been your most successful contracting techniques? How has their effectiveness been measured? Please comment on cost, time, quality, insurance, bonding, and stakeholder issues.

II. Design-Build

General Definitions

II-1. Design-build (DB) is a delivery method that utilizes only one contract for both design and construction. These contracts are typically awarded on a lump sum basis. Is your country utilizing the DB delivery method? If not, please skip to the next section.

II-2. Are additional services such as maintenance or operation of the facility being utilized with DB delivery?

Project and Delivery Team Selection

II-3. What criteria do your country use to select projects for the DB delivery method?

II-4. What process does your country use to select the DB teams? Does your country use a prequalification or shortlisting process?

Scope Definition and Risk Allocation

II-5. What has been the most successful mix of percentage of plans complete, agency intervention, warranty usage, and incentives?

II-6. Is the design "approved" or "reviewed" by the owner agency as the project progresses? In other words, at what point in the process does the owner commit to the final design?

II-7. How are changes to the contract after award addressed in a DB project?

II-8. How are risks allocated for third-party utilities, right-of-way, and environmental issues?

Quality Assurance and Quality Control

II-9. How does your country ensure high design and construction quality of the finished product?

II-10. What is the inspection and testing process for the owner? For the design-builder?

II-11. Does the owner have the authority to stop work at any time for quality concerns? If so, what are the contract ramifications of this?

Global Effects

II-12. What have been your most successful DB projects and processes? How has their effectiveness been measured? Please comment on cost, time, quality, insurance, bonding, and stakeholder issues.

III. Alternative Financing

General Definitions

III-1. Please describe your traditional method of project financing.

III-2. What alternative financing techniques is your country employing?

III-3. If revenue financing, what is the structure and how does it fit into the project development process?

Public-Private Partnerships

III-4. Do your country's laws allow for public-private partnerships? If so, what were the major obstacles to initiating these partnerships?

III-5. In what types of public-private arrangements does your country engage?

III-6. Does your government provide any loan guarantees, lines of credit, or tax incentives to sponsors of large surface transportation projects? Are these national or regional?

Project and Partnership Selection

III-7. What criteria do your country use to select projects for alternative financing methods?

III-8. How are private partners selected? Is there competition involved in creating these partnerships?

Scope Definition and Risk Allocation

III-9. What level of project oversight is employed when using public-private relationships?

III-10. How does your country control the associated risks when using nontraditional methods?

III-11. What is the appropriate length of time that the partnership should be in effect?

Global Effects

III-12. What have been your most successful alternative financing arrangements? How has their effectiveness been measured? Please comment on cost, time, quality, and stakeholder issues.

IV. Performance Contracting

General Definitions

IV-1. Is your country utilizing performance contracting in lieu of low-bid methods? If so, please describe the specific performance contracting methods.

IV-2. Describe the process for developing performance criteria associated with performance methods.

Project and Contractor Selection

IV-3. What criteria do your country use to select projects for performance contracting?

IV-4. What are the performance criteria that your country uses for selecting contractors (i.e., cost, technical excellence, management capability, past performance, personnel qualifications, financial capacity, etc.)?

IV-5. Are any unique dispute-resolution techniques utilized with performance contracting?

Warranties

IV-6. What types of work are being warrantied in your country? Please describe both individual items and entire projects.

IV-7. What are the lengths of typical warranties? How are unique or unexpected loads accounted for on extended warranties?

IV-8. Is the contractor given more control over design and construction details when using warranties?

IV-9. How does your country inspect the warranted work during the warranty period?

IV-10. Who is responsible for routine maintenance and emergency maintenance during the warranty period?

IV-11. Is liability for latent defects cut off at the end of the warranty term? What limitations on liability are included in the contract?

Quality Assurance and Quality Control

IV-12. Who is responsible for quality control and quality assurance in performance contracting arrangements?

IV-13. How does performance contracting affect inspection requirements?

IV-14. What policies are in place to ensure that contractors are being measured for their performance and not penalized for inadequate designs developed by others?

Global Effects

IV-15. What have been your most successful performance contracting arrangements? How has their effectiveness been measured? Please comment on time, cost, quality, and stakeholder issues.

V. Payment Methods

General Definitions

V-1. Payment for unit of work as bid by the unit price is the traditional method of payment for work in the United States, but this method is not appropriate for some alternative delivery methods (design-build or other innovative contracts). What other processes is your country utilizing to improve the payment process?

Contract and Payment Measurement Techniques

V-2. To what extent does your country utilize lump sum bidding? What are the associated payment methods?

V-3. Does your country use actual (reimbursable) cost with a guaranteed maximum price type contracting? If so, what is the process for administering the contract and payments?

V-4. In what cases do the contractor, rather than the owner, conduct the measurements for payments?

V-5. How do you document and pay for traffic control on construction projects?

Incentive Payments

V-6. In the United States we are employing incentive payments to increase the quality of construction. Does your country utilize incentive payments?

V-7. On what types of work is your country employing incentives and what are the associated measures upon which the incentives are based?

V-8. How does your country determine the amount of the incentive payment such that it will motivate the contractor yet be cost-effective to the owner agency? Are these amounts capped at some maximum value?

Global Effects

V-9. What have been your most successful payment method arrangements? How has their effectiveness been measured? Please comment on time, cost, quality, and stakeholder issues.

VI. Asset Management

General Definitions

VI-1. Asset management involves maintaining and operating existing infrastructure. Please describe the asset management process in your country.

VI-2. What innovative contracting techniques are being utilized for asset management? Are any of these techniques tied to new technologies (sensors, databases, etc.)?

Maintenance and Operation

VI-3. What types of work are utilizing innovative maintenance and operation techniques?

VI-4. What provisions ensure that maintenance and repairs are performed? To what extent are bonds utilized?

VI-5. How is competition promoted on maintenance and operation contracts?

VI-6. What is the typical length of operation and maintenance contracts?

VI-7. Are there concerns over "risk shifting" to the industry with long-term detrimental effects on smaller firms and their ability to continue working?

VI-8. What type of referee system is used when differences of opinion arise as to responsibility for failures?

Global Effects

VI-9. What have been your most successful asset-management techniques? How has their effectiveness been measured? Please comment on time, cost, quality, insurance, bonding, and stakeholder issues.

www.ingramcontent.com/pod-product-compliance
Lightning Source LLC
Chambersburg PA
CBHW080303180526
45167CB00006B/2647
* 9 7 8 1 4 9 3 6 2 0 4 4 9 *